AF388982

Integration and Control of Distributed Renewable Energy Resources

Integration and Control of Distributed Renewable Energy Resources

Editor

Hamidreza Nazaripouya

MDPI • Basel • Beijing • Wuhan • Barcelona • Belgrade • Manchester • Tokyo • Cluj • Tianjin

Editor
Hamidreza Nazaripouya
Oklahoma State University
USA

Editorial Office
MDPI
St. Alban-Anlage 66
4052 Basel, Switzerland

This is a reprint of articles from the Special Issue published online in the open access journal *Clean Technologies* (ISSN 2571-8797) (available at: https://www.mdpi.com/journal/cleantechnol/special_issues/renewable_energy_resources).

For citation purposes, cite each article independently as indicated on the article page online and as indicated below:

LastName, A.A.; LastName, B.B.; LastName, C.C. Article Title. *Journal Name* **Year**, *Volume Number*, Page Range.

ISBN 978-3-0365-3689-7 (Hbk)
ISBN 978-3-0365-3690-3 (PDF)

Contents

About the Editor

Hamidreza Nazaripouya received his B.S. degree in electrical engineering from the University of Tehran, Iran, his M.Sc. degree in power electronics from the Sharif University of Technology, Tehran, Iran, and his M.Sc. degree in power systems from Louisiana State University, LA, USA. He obtained his Ph.D. degree from the University of California, Los Angeles (UCLA). He is currently an Assistant Professor at Oklahoma State University, OK, USA and an Assistant Adjunct Professor at the University of California, Riverside (UCR), CA, USA. His research interests include cyber–physical power systems; the automation, sustainability, and resilience of power systems; the control and integration of renewable energy, energy storage, electric vehicles, and power electronics. He holds U.S. patents in the control of energy storage systems. His patented technology won the NSF Grant Award with him as the Entrepreneurial Lead. He also received IEEE SFV Section Rookie of the Year Award, IEEE IAS and PES Presentation awards, and the UC Dissertation-Year Fellowship Award.

clean technologies

MDPI

Editorial

Integration and Control of Distributed Renewable Energy Resources

Hamidreza Nazaripouya [1,2]

1   School of Electrical and Computer Engineering, Oklahoma State University, Stillwater, OK 74078, USA; hanazar@okstate.edu
2   Department of Electrical and Computer Engineering, University of California, Riverside (UCR), Riverside, CA 92507, USA

Citation: Nazaripouya, H. Integration and Control of Distributed Renewable Energy Resources. *Clean Technol.* 2022, 4, 149–152. https://doi.org/10.3390/cleantechnol4010010

Received: 7 February 2022
Accepted: 24 February 2022
Published: 1 March 2022

Publisher's Note: MDPI stays neutral with regard to jurisdictional claims in published maps and institutional affiliations.

1. Introduction

The increase in the population growth rate and the motivation to overcome issues such as environmental concerns and air pollution have made distributed renewable energy resources (DRER) the most popular option for providing the required energy. In addition, DRER, by making technologies such as microgrid feasible, facilitate the expansion of the power network to disadvantaged communities as well as remote and rural areas, which are not connected to the main grid. At the same time, they can improve the operation, resilience, and efficiency of the power grid by providing extra flexibility, local energy supply, and lower energy loss [1,2].

However, there are still several technical and non-technical challenges regarding interconnection of renewables into the grid [3]. The current power grid is aging and faces a future for which it was not designed. For instance, with the widespread proliferation of rooftop solar, distribution utilities have consumers who become "prosumers" as they consume and produce electricity at the same time [4]. However, the traditional electric power grid has been designed for power flow in one direction. As the penetration of distributed renewable energy resources such as solar or wind increases, it is anticipated that in some feeders, the flow of power changes to the "reverse direction" [5]. In addition, the power fluctuations of DRER adversely affect the operation of power systems [6]. From the grid perspective, technical concerns associated with the high penetration of renewables with intermittent nature include energy management [7,8], forecasting [9,10], power quality (e.g., variations of voltage and harmonic issues) [5], grid stability [11], reliability, and protection issues.

Moreover, still many aspects of DRER deployment are nascent, for which there are no standards or accepted best practices [12]. For example, the deployment of DRER increases the number of grid-edge devices that are typically equipped with communication and control interfaces for remote access and control. These emerging grid-edge devices introduce new vulnerabilities to the gid when it comes to cyber-physical attacks. From the DRER perspective, the technical challenges range over a wide spectrum of topics including technoeconomic feasibility analysis, technology selection, software/hardware design, modeling, planning, sizing and placement, grid support capability, cyber-physical security, and interoperability at the component level and system level.

On the other hand, the non-technical challenges can be classified in three categories—regulatory issues, social issues, economic issues. Regarding regulatory issues, impotent policies, lack of standards and regulations, complex administrative procedures, and sometimes the gap between policy targets and reality, are some examples. Social issues include, but not limited to, lack of public awareness and information, lack of technically skilled labor, and adverse impacts on other/alternative sources of income. Competition with fossil fuel, lack of effective incentives and subsidies, high capital cost, and challenges in securing financing are some existing economic issues [13].

To overcome the existing technical and non-technical challenges, considerations should be taken into account for integration and control of DRER. Understanding these considerations across a wide range of topics associated with DRER deployment as well as becoming familiar with emerging practices and solutions pave the path to achieving a high level of renewable penetration in power systems.

2. A Short Review of the Contributions in This Issue

This Special Issue is focused on "Integration and Control of Distributed Renewable Energy Resources" and aims to identify some of the considerations and new practices for interconnection and operation of DRER in different applications. Topics addressed in this Special Issue include technology selection of photovoltaics (PV) for optimal operation of solar-powered electric vehicle charging facilities, design of small-scale wind turbines for residential applications, hosting capacity analysis of solar PV systems in distribution grids, optimal allocation of DRER in a distribution network, resilient operation of distributed energy resources in a distribution grid, hardware control of a hybrid wind–PV-battery power-generation system, and finally control of single-phase inverter-interfaced DRER to achieve a desired interphase power flow/routing.

In this issue, a total of seven papers were published, covering different aspects of integration and control of DRER including technology selection, design, techno-economic analysis, planning, management, and operation.

Four papers are focused on control and optimization of DRER operation. The first one developed a composite control strategy for off-grid operation of a hybrid wind–solar-battery–diesel power-generation system [14]. Some features of the proposed control include speed sensorless operation of a variable-speed wind turbine, stable operation of the system during disturbances, enhanced power quality at the point of connection under nonlinear loads, and extraction of maximum power from PV without using any maximum power point tracking (MPPT) algorithm [14]. The second paper proposed an interphase power flow (IPPF) control for line-to-line single-phase power electronic-interfaced DRER [15]. The IPPF controller allows single-phase elements to route active power between phases, improving system operation and flexibility. The IPPF control was also applied to a utility distribution circuit, which led to reductions in system voltage unbalance and losses through balancing the active power at the feeder head. The third one [16] considered four different PV panel technologies and optimized the operation of Solar PV systems in an electric vehicle charging center. The profit of the charging facility was maximized considering solar radiation uncertainty and different behaviors of EV owners as well as several weather conditions. Finally, a model-based predictive control (MPC) strategy was proposed for power flow management in a power distribution system with PV power generation [17]. The optimization problem is subject to voltage constraints and assets' operational restrictions. A forecasting module based on a Gaussian process regression model was constructed to predict the global horizontal irradiance (GHI), grid load, and water demand. It was shown that the proposed MPC strategy is resilient to errors of the forecasting module. A min–max problem was added on top of the main optimization problem to minimize voltage overflow, which also enhances the resilience of the strategy.

The investigation into optimized allocation of distributed energy resources (DERs) in distribution systems with the goal of minimizing the system loss was pursued in [18]. A hybrid optimization approach, composed of the tunicate swarm algorithm (TSA) and the sine–cosine algorithm (SCA), was introduced to identify the best size and location of DERs in the system. The optimization process is performed for DERs in three different modes: active power production (P-mode), reactive power production (Q mode), and active and reactive power production (PQ-mode). Authors in [19] conducted an investment analysis for integration of a small-scale wind turbine into residential homes. Their studies show that relying only on software tools may mislead investors during the decision-making process, as in most of these tools, installation costs, maintenance costs, net metering options, and taxation schemes are not included. The focus in [20] is on assessing the threshold of PV penetration, known as hosting capacity (HC) of PV integration, in low-voltage distribution systems. To this end, the HC values of three types of networks in rural, suburban, and urban regions for different HC reference definitions were compared. The comparison was made under balanced and unbalanced PV deployment scenarios and different load conditions. Monte Carlo (MC) simulations and stochastic analysis were utilized to include the effect of PV power intermittency and varying loading conditions in the assessment.

3. Conclusions

In the 21st-century technological race for power-grid modernization, DRER provide an opportunity to help shape the future of our nation's energy supply while improving the reliability, resilience, affordability, flexibility, and security of the electric power grid, and providing a chance to solve the corresponding economic, environmental, and social problems. The diverse contributions in this Special Issue make it evident that the deployment of distributed renewable energy resources is a vast topic and has many aspects to be studied in detail. As the demand and interest for producing clean energies are growing rapidly, there is a need to understand considerations for interconnection and control of distributed renewable resources. This Special Issue aimed to shed some light on some key issues as well as potential solutions in DRER integration and control.

Funding: This research received no external funding.

Institutional Review Board Statement: Not applicable.

Informed Consent Statement: Not applicable.

Data Availability Statement: Not applicable.

Acknowledgments: The Guest Editor sincerely appreciates all the authors in this Special Issue for their efforts and valuable contributions.

Conflicts of Interest: The authors declare no conflict of interest. Additionally, the funders had no role in the design of the study; in the collection, analyses, or interpretation of data; in the writing of the manuscript, or in the decision to publish the results.

References

1. Nazaripouya, H.; Pota, H.R.; Chu, C.-C.; Gadh, R. Real-Time Model-Free Coordination of Active and Reactive Powers of Distributed Energy Resources to Improve Voltage Regulation in Distribution Systems. *IEEE Trans. Sustain. Energy* **2020**, *11*, 1483–1494. [CrossRef]
2. Nazaripouya, H. Grid Resilience and Distributed Energy Storage Systems. *IEEE Smart Grid Newsl. Smart Grid Appl. Enhanc.* **2019**, *17*.
3. Nazaripouya, H. Integration and Control of Battery Energy Storage and Inverter to Provide Grid Services. Ph.D. Thesis, University of California, Los Angeles, CA, USA, 2017.
4. Azar, A.G.; Nazaripouya, H.; Khaki, B.; Chu, C.C.; Gadh, R.; Jacobsen, R.H. A Non-Cooperative Framework for Coordinating a Neighborhood of Distributed Prosumers. *IEEE Trans. Ind. Inform.* **2018**, *15*, 2523–2534. [CrossRef]
5. Nazaripouya, H.; Wang, Y.; Chu, P.; Pota, H.R.; Gadh, R. Optimal Sizing and Placement of Battery Energy Storage in Distribution System Based on Solar Size for Voltage Regulation. In Proceedings of the 2015 IEEE PES General Meeting, Denver, CO, USA, 26–30 July 2015.
6. Nazaripouya, H.; Chu, C.-C.; Pota, H.R.; Gadh, R. Battery Energy Storage System Control for Intermittency Smoothing Using an Optimized Two-Stage Filter. *IEEE Trans. Sustain. Energy* **2018**, *9*, 664–675. [CrossRef]
7. Wang, Y.; Nazaripouya, H. Demand-Side Management in Micro-Grids and Distribution Systems: Handling System Uncertainties and Scalabilities. In *Classical and Recent Aspects of Power System Optimization*; Zobaa, A.F., Aleem, S.H., Abdelaziz, A.Y., Eds.; Elsevier Inc.: London, UK, 2018.
8. Wang, Y.; Wang, B.; Zhang, T.; Nazaripouya, H.; Chu, C.C.; Gadh, R. Optimal energy management for Microgrid with stationary and mobile storages. In Proceedings of the 2016 IEEE/PES Transmission and Distribution Conference and Exposition (T&D), Dallas, TX, USA, 3–5 May 2016; pp. 1–5.
9. Nazaripouya, H.; Wang, B.; Wang, Y.; Chu, P.; Pota, H.R.; Gadh, R. Univariate time series prediction of solar power using a hybrid wavelet-ARMA-NARX prediction method. In Proceedings of the 2016 IEEE/PES Transmission and Distribution Conference and Exposition (T&D), Dallas, TX, USA, 3–5 May 2016.
10. Majidpour, M.; Nazaripouya, H.; Chu, C.; Pota, H.R.; Gadh, R. Fast Univariate Time Series Prediction of Solar Power for Real-Time Control of Energy Storage System. *Forecasting* **2018**, *1*, 107–120. [CrossRef]

11. Saberi, H.; Nazaripouya, H.; Mehraeen, S. Implementation of a Stable Solar-Powered Microgrid Testbed for Remote Applications. *Sustainability* **2021**, *13*, 2707. [CrossRef]

12. Horowitz, K.; Peterson, Z.; Coddington, M.; Ding, F.; Sigrin, B.; Saleem, D.; Schroeder, C. *An Overview of Distributed Energy Resource (DER) Interconnection: Current Practices and Emerging Solutions*; National Renewable Energy Laboratory (NREL): Golden, CO, USA, 2019.

13. Seetharaman; Moorthy, K.; Patwa, N.; Saravanan; Gupta, Y. Breaking Barriers in Deployment of Renewable Energy. *Heliyon* **2019**, *5*, e01166. [CrossRef] [PubMed]

14. Rezkallah, M.; Ibrahim, H.; Dubuisson, F.; Chandra, A.; Singh, S.; Singh, B.; Issa, M. Hardware Implementation of Composite Control Strategy for Wind-PV-Battery Hybrid Off-Grid Power Generation System. *Clean Technol.* **2021**, *3*, 821–843. [CrossRef]

15. Siratarnsophon, P.; Cunha, V.C.; Barry, N.G.; Santoso, S. Interphase Power Flow Control via Single-Phase Elements in Distribution Systems. *Clean Technol.* **2021**, *3*, 37–58. [CrossRef]

16. Farahmand, M.Z.; Javadi, S.; Sadati, S.M.; Laaksonen, H.; Shafie-khah, M. Optimal Operation of Solar Powered Electric Vehicle Parking Lots Considering Different Photovoltaic Technologies. *Clean Technol.* **2021**, *3*, 503–518. [CrossRef]

17. Dkhili, N.; Eynard, J.; Thil, S.; Grieu, S. Resilient Predictive Control Coupled with aWorst-Case Scenario Approach for a Distributed-Generation-Rich Power Distribution Grid. *Clean Technol.* **2021**, *3*, 629–655. [CrossRef]

18. Awad, A.; Abdel-Mawgoud, H.; Kamel, S.; Ibrahim, A.A.; Jurado, F. Developing a Hybrid Optimization Algorithm for Optimal Allocation of Renewable DGs in Distribution Network. *Clean Technol.* **2021**, *3*, 409–423. [CrossRef]

19. Ribbing, S.; Xydis, G. Renewable Energy at Home: A Look into Purchasing a Wind Turbine for Home Use—The Cost of Blindly Relying on One Tool in Decision Making. *Clean Technol.* **2021**, *3*, 299–310. [CrossRef]

20. Fatima, S.; Püvi, V.; Lehtonen, M. Comparison of Different References When Assessing PV HC in Distribution Networks. *Clean Technol.* **2021**, *3*, 123–137. [CrossRef]

Article

Hardware Implementation of Composite Control Strategy for Wind-PV-Battery Hybrid Off-Grid Power Generation System

Miloud Rezkallah [1,2,*], Hussein Ibrahim [1], Félix Dubuisson [2], Ambrish Chandra [2], Sanjeev Singh [3], Bhim Singh [4] and Mohamad Issa [5]

[1] CR2ie Sept-Îles, 175 Rue de la Vérendrye, Sept-Îles, QC G4R 5B7, Canada; Hussein.Ibrahim@cegepsi.ca
[2] Department of Electrical Engineering, École de Techonologie Superieure, 1100 Notre-Dame Montréal, Montreal, QC H3C1K3, Canada; felix.dubuisson.1@ens.etsmtl.ca (F.D.); Ambrish.chandra@etsmt.ca (A.C.)
[3] Electrical Engineering Department, Maulana Azad National Institute of Technology, Bhopal 462051, India; sschauhan@manit.ac.in
[4] Electrical Engineering Department, Indian Institute of Technology Delhi, New Delhi 110016, India; bsingh@ee.iitd.ac.in
[5] Quebec Maritime Institute, Rimouski, QC G5L 4B4, Canada; missa@imq.qc.ca
* Correspondence: miloud.rezkallah@itmi.ca or miloud.rezkallah@etsmtl.ca

Citation: Rezkallah, M.; Ibrahim, H.; Dubuisson, F.; Chandra, A.; Singh, S.; Singh, B.; Issa, M. Hardware Implementation of Composite Control Strategy for Wind-PV-Battery Hybrid Off-Grid Power Generation System. *Clean Technol.* 2021, 3, 821–843. https://doi.org/10.3390/cleantechnol3040048

Academic Editor: Hamidreza Nazaripouya

Received: 31 May 2021
Accepted: 4 November 2021
Published: 16 November 2021

Publisher's Note: MDPI stays neutral with regard to jurisdictional claims in published maps and institutional affiliations.

Abstract: In this paper, a composite control strategy for improved off-grid configuration based on photovoltaic (PV array), a wind turbine (WT), and a diesel engine (DE) generator to achieve high performance while supplying nonlinear loads is investigated. To operate the WT efficiently under variable speed conditions and to obtain accurate and fast convergence to the maximum global operating point without a speed sensor, an iterative interpolation method is integrated with the perturbation and observation (P&O) technique. To ensure the balance of power in the system and to achieve the maximum power from the PV array without using any maximum power point tracking (MPPT) method, and ensuring stable operation during the disturbance, a double-loop control strategy for a two-switches buck-boost converter is developed. Furthermore, to protect the synchronous generator of the diesel generator (DG) from the 5th and 7th order-harmonics created by the connected nonlinear loads and to solve the issue of the filter resonance, the interfacing three-phase inverter is controlled using an improved synchronous-reference frame algorithm (SRF) with virtual impedance active damping. The presented work demonstrates effective and efficient control along with improved performance and cost-effective option as compared to the similar works reported in the literature. The performance of the presented off-grid configuration and its developed composite control strategy are tested using MATLAB/Simulink and validated through small-scale hardware prototyping.

Keywords: off-grid system; composite control strategy; solar photovoltaic panel; wind turbine; diesel generator; energy storage system (ESS); synchronous machine (SM); permanent magnet brushless DC machine (PMBLDCM); power quality improvement

1. Introduction

The hybrid off-grid power generation system based on renewable energy sources (RES), such as a wind turbine (WT), photovoltaics (PVs array) a non-renewable diesel generator (DG), and an energy storage system (ESS), has demonstrated its capability to provide uninterrupted and clean energy to the connected electrical local loads at low cost [1–3]. Many off-grid configurations based on the hybridization of different RESs, and their control strategies, are detailed in [4,5]. In all these off-grid configurations, non-renewable energy sources (N-RESs) are suggested as reliable energy sources to compensate for the intermittency of RES and always ensure uninterruptible power supply to the connected local loads. Detailed comparisons on off-grid configurations, control design, and application are given in [6–8]. DG, which consists of a diesel engine (DE) and electrical machine in a hybrid off-grid system, is employed as backup ES and is connected directly to the point of common coupling (PCC) to fulfill the connected linear and nonlinear loads, as detailed in all off-grid configurations presented in [9]. According to the detailed study, realized by [9] on the induction machine, the harmonics generated by nonlinear loads can affect the performance of the electrical and mechanical parts of the DG, and in [10], the authors have found that 5th and 7th harmonics have a significant impact on the performance of the synchronous machine. To prevent this issue and to improve the operational effectiveness of DG, in [10], the 5th and 7th harmonics are eliminated by controlling the interfacing inverter as a shunt active filter using multifunction control algorithm based on PRC controllers, while showing satisfactory performance under all types of loads. In the same context, the authors in [11] have suggested a passive harmonic filter to mitigate selective harmonics. Compared to an active power filter (APF), the passive power filter (PPF) is less complex and easy to design, but the active filter possesses a high value of the quality factor, which is suggested for applications where many energy sources are connected to the PCC, as in this case. Many research works are reported in the literature on the control and design of the APF [12–14]. Unfortunately, all the concepts given in [12–14] are dedicated to APF application, which has only one task in a hybrid off-grid application. So, the multitask option is preferred as detailed in [10], where frequency and voltage regulation at the PCC are taken into consideration. In [15–17], the synchronous reference frame technique (SRF) is proposed for standalone and grid-connected systems. The obtained results under the presence of nonlinear loads show satisfactory performance. Unfortunately, the performance of this technique for under-voltage and frequency variation has not been presented. In [18], sliding mode control (SMC) is suggested to maintain constant PCC voltage and frequency, and to improve the power quality, simultaneously. The authors have succeeded to validate the proposed control for the off-grid system in real-time, and the obtained results show satisfactory performance. Compared to the SRF technique proposed in [15–17], or the instantaneous power theory, SMC is complex and requires an accurate technique to select the optimal SM gains (β_1 and β_2) to achieve high performance.

The second non-renewable energy source (N-RES) is the energy storage system (ESS), which is considered as the key to the stable operation of the off-grid system, especially if it contains RESs. Generally, the ESS is connected to the common DC bus through a buck-boost converter [18,19] and is controlled to balance the power in the system and compensate for the wind and solar power intermittency and variability. In [20], SMC for DC voltage and battery current regulation is proposed, and in [21], a control scheme based on the state of charge of the battery (SoC%) is employed to charge and discharge lithium-ion battery through a three-phase DC-AC converter. For both studies detailed in [20,21], the authors have succeeded to implement their control strategies in real-time, and the obtained results show the capability of the ESS to maintain the stable operation of the hybrid off-grid system. Unlike the SMC suggested in [20], the strategy based on the SoC%, which is suggested in [21], is simple and performs well under sudden variations of solar irradiation and load change.

With regard to the RES integration in off-grid systems and their output power optimization, authors in [4,5] have proposed a hybrid AC/DC configuration. For output power optimization, perturbation and observation (P&O) technique is widely employed for WT as well as PV array due to its simplicity [22]. According to [23], the P&O technique cannot perform well under a sudden change in solar irradiation or under partial shading. To improve their performance, in [24], P&O is reinforced by the adaptive algorithm to adjust the reference step size, and in [25], the resistance P&O is combined with adaptive resistance control to achieve high performance from the PV panel under sudden solar irradiation change. In [24,25], the objectives are achieved but the system complexity is increased compared to the conventional P&O technique.

For improvement in the performance of the off-grid configurations available and reported in [26–28], the following solutions are presented in this paper as significant contributions,

1. Minimizing the number of power converters to reduce the hardware complexity and increase the system efficiency,
2. Development of an indirect control for the buck-boost converter to realize many tasks such as achieving high performance from PV without using any MPPT algorithm, facilitating the bidirectional power flow between the ESS and PCC, and ensuring stable operation during the disturbance,
3. Effective and efficient, mechanical-speed sensorless operation of variable-speed WT-based permanent magnet brushless DC generator (PMBLDCG) using hybridization of the root-finding algorithm (secant method) with P&O technique,
4. Reinforcement of the SRF based control with virtual impedance active damping to improve the power quality at the PCC while eliminating the 5th and 7th order harmonics, along with the prevention of the 6th order-harmonic generation in the rotor of the synchronous generator (SG), as well as to solve the issue of filter resonance.

2. System Configuration and Operation

Figure 1 demonstrates an improved version of the off-grid configurations proposed in [26–28]. It consists of WT-driven variable speed PMBLDCG, PV, ESS, and DE-driven fixed speed synchronous generator (SG). To achieve high performance from WT without using any mechanical speed sensor, the stator terminals of the PMBLDCG are connected to the common DC bus through a three-phase diode bridge and DC-DC boost converter, which is controlled using the root-finding algorithm (secant method) with the P&O technique. To achieve high performance from PV without losses in power converter, their terminals are connected directly to the common DC bus. The ESS is connected to the common DC bus through a DC-DC buck-boost converter, which is controlled using indirect method-based control to achieve many tasks simultaneously. The WT, PV, and ESS are connected to the PCC through a three-phase interfacing inverter, which is interfaced with an LCL passive filter. To protect the DG, which consists of DE-driven fixed speed SG against the 5th and 7th order harmonics and prevent the generation of the 6th order-harmonic in the rotor of the SG, and avoid the filter resonance, SRF control with virtual impedance active damping is employed.

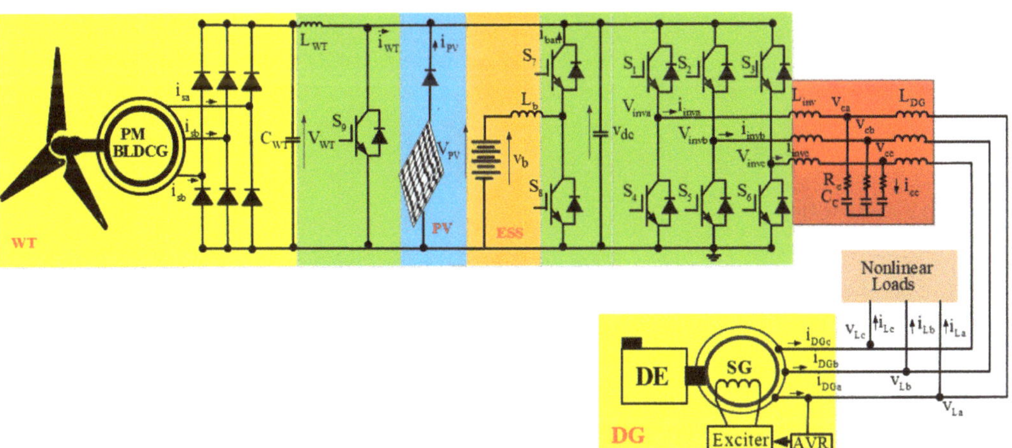

Figure 1. Improved hybrid wind-PV off-grid configuration under study.

The operation modes of the off-grid system are decided by generated and consumed powers as presented in Table 1. There are four operating modes with specific conditions, such as the state of charge of the battery (SoC%), the generated power from ESs, and the load power demand for the decision of the operating mode. In this study, the operating modes are decided based on the sum of the generated power from WT (P_{WT}), PV (P_{PV}), and DG (P_{DG}), depending on if it is greater, equal, or less than load power demand (P_L), and also based on the SoC% of ESS.

Table 1. Operation modes of the system.

Mode	Conditions	ES	State of ESS
Mode 1	$P_{PV} + P_{WT} + P_{DG} < P_L$ SoC% > 50%	WT, PV&DG	discharging
Mode 2	$P_{PV} + P_{WT} + P_{DG} > P_L$ SoC% < 50%	WT, PV&DG	charging
Mode 3	$P_{PV} + P_{WT} + P_{DG} > P_L$ SoC% < 100%	WT, PV&DG	charging
Mode 4	SoC% = 100% $P_{pv} + P_{WT} + P_{DG} > P_L$ $P_{pv} + P_{WT} + P_{DG} = P_L$	WT, PV&DG	Stop charging

3. Developed Composite Control Strategy

In this section, the developed control strategies for MPPT of the WT and PV panel, charging and discharging the ESS, and energy management, as well as power quality improvement at the PCC, are detailed.

3.1. Control of DC-DC Boost Converter on WT Side

Figure 2 shows the developed enhanced control strategy to achieve MPPT from WT without using a mechanical speed sensor. P&O algorithm, which is detailed in [29], is combined with an iterative interpolation method called secant method with variable step for V_{WT} to provide accurate and fast convergence to the optimum operating point during sudden wind speed change. The fundamental equation of the secant method is described in [30–33].

$$x_{(n+1)} = x_n - \left(F(x_n) \frac{\left(x_n - x_{(n-1)} \right)}{F(x_n) - F\left(x_{(n-1)} \right)} \right) \tag{1}$$

where x_n the initial value of x and $F(x_n)$ represents the value of the function at x_n, which is described as,

$$F(x_n) = F\left(V_{(WT(n))} \right) = \frac{dP_{WT}}{dV_{WT}} \tag{2}$$

where V_{WT} and P_{WT} represent the output DC voltage and the generated power of WT.

$$F\left(V_{(WT(n))} \right) = \frac{P_{WT(n)} - P_{WT(n-1)}}{V_{WT(n)} - V_{WT(n-1)}} = \nabla(n) \tag{3}$$

where $(\nabla(n))$ is the gradient of (n).

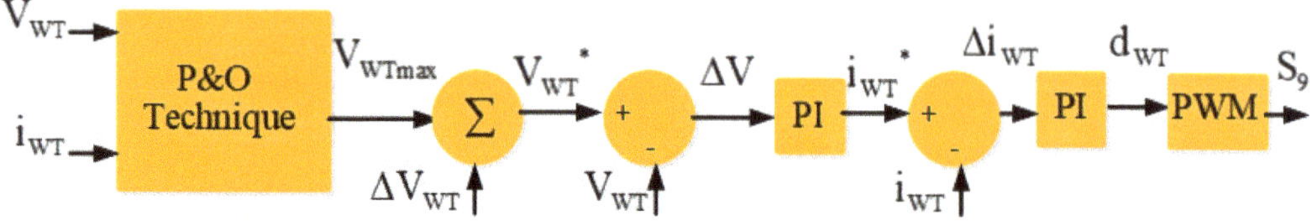

Figure 2. P&O-based secant method and variable step for DC-DC-boost converter.

From (2) and (3), the step of V_{WT} variation (ΔV_{WT}) is obtained as,

$$\Delta V_{WT(n)} = \nabla(n)\frac{V_{WT(n)} - V_{WT(n-1)}}{\nabla(n) - \nabla(n-1)} \tag{4}$$

The reference voltage V^*_{WT} is obtained as,

$$V^*_{WT} = V_{WTmax(n)} + \Delta V_{WT} \tag{5}$$

where $V_{WTmax(n)}$ is the maximum voltage obtained using the P&O technique as detailed in Figure 2.

The reference V^*_{WT} is compared with sensed DC voltage at the output of the three-phase diode bridge. The error of voltage ΔV is fed to the PI voltage controller. The obtained signal from the outer loop control represents the DC WT current reference (i^*_{WT}) as

$$i^*_{WT} = K_{p1}\Delta V + K_{i1} \int \Delta V \mathrm{dt} \tag{6}$$

where K_{p1} and K_{i1} represent the proportional and integral gains of the outer control loop, and ΔV is the WT DC voltage error value.

The DC WT current reference is compared with the measured DC WT current (i_{WT}), and the current error (Δi_{WT}) is fed to PI current controller to get control signal (d_{WT}) as

$$d_{WT} = K_{p1}\Delta i_{WT} + K_{i1} \int \Delta i_{WT}\mathrm{dt} \tag{7}$$

where K_{p2} and K_{i2} represent the proportional and integral gains of the inner control loop, and Δi_{WT} is the WT-DC error value.

This output signal of the PI controller is fed to the PWM controller to get the switching control of the switch (S_9) of the boost converter for the WT side.

3.2. Control of DC-DC Buck-Boost Converter for ESS Side

Figure 3 shows the block diagram of the developed indirect control of the DC-DC buck-boost converter. Based on the control of the DC voltage of the common DC bus, one can easily get the maximum power from the PV panel without using any MPPT technique and controlled power converter by connecting many PVs in series to make the output PV panels voltage equal to the common DC bus voltage. So, by controlling the common DC bus voltage (V_{dc}), one balances the power in the hybrid off-grid system and indirectly gets the MPP from the PV array. As shown in Figure 3, two control loops are employed to control the DC-DC buck-boost converter for the ESS side. The error of the common DC bus voltage (ΔV_{dc}) is calculated as

$$\Delta V_{dc} = V^*_{dc} - V_{dc} \tag{8}$$

where V^*_{dc} and V_{dc} represent the common reference DC bus voltage and its measured signal.

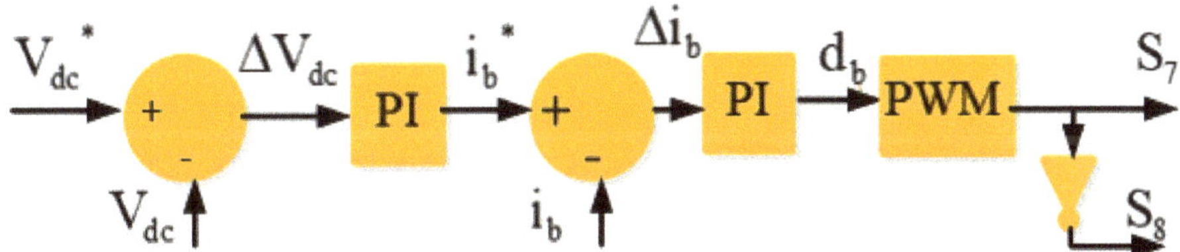

Figure 3. Indirect control for the DC-DC buck-boost converter for the ESS side.

The obtained error value of the common DC bus voltage (ΔV_{dc}) is fed to the PI voltage controller to get the ESS current reference (i_b^*) as

$$i_b^* = K_{p3}\Delta V_{dc} + K_{i3}\int \Delta V_{dc}\mathrm{dt} \tag{9}$$

where K_{p3} and K_{i3} represent the proportional and integral gains of the outer control loop, and ΔV_{dc} is the error value of the common DC bus voltage.

The DC-WT current reference is compared with the measured value (i_b), and the current error value (Δi_b) is fed to PI current controller of the inner control loop to get control signal (d_b) as

$$d_b = K_{p4}\Delta i_b + K_{i4}\int \Delta i_b\mathrm{dt} \tag{10}$$

where K_{p4} and K_{i4} represent the proportional and integral gains of the inner control loop.

The output signal of the PI controller of the inner control loop is fed to the PWM controller to get the switching control of switches (S_7 and S_8) of the DC-DC buck-boost converter for the ESS side.

3.3. SRF Control with Virtual Impedance-Based Active Damping

Figure 4 shows the block diagram of synchronous reference frame (SRF) control based on virtual impedance active damping. The theory of SRF control as detailed in [14] is reinforced by virtual impedance-based active damping to dampen the LCL resonance. The measured load currents (i_{La}, i_{Lb}, i_{Lc}) and PCC voltages (v_{La}, v_{Lb}, v_{Lc}) are converted into the d-q-o frame using Park's transformation as,

$$\begin{bmatrix} i_{Ld} \\ i_{Lq} \\ i_{Lo} \end{bmatrix} = \frac{2}{3}\begin{bmatrix} \cos\omega t & -\sin\omega t & \frac{1}{2} \\ \cos\left(\omega t - \frac{2\pi}{3}\right) & -\sin\left(\omega t - \frac{2\pi}{3}\right) & \frac{1}{2} \\ \cos\left(\omega t + \frac{2\pi}{3}\right) & \sin\left(\omega t + \frac{2\pi}{3}\right) & \frac{1}{2} \end{bmatrix}\begin{bmatrix} i_{La} \\ i_{Lb} \\ i_{Lc} \end{bmatrix} \tag{11}$$

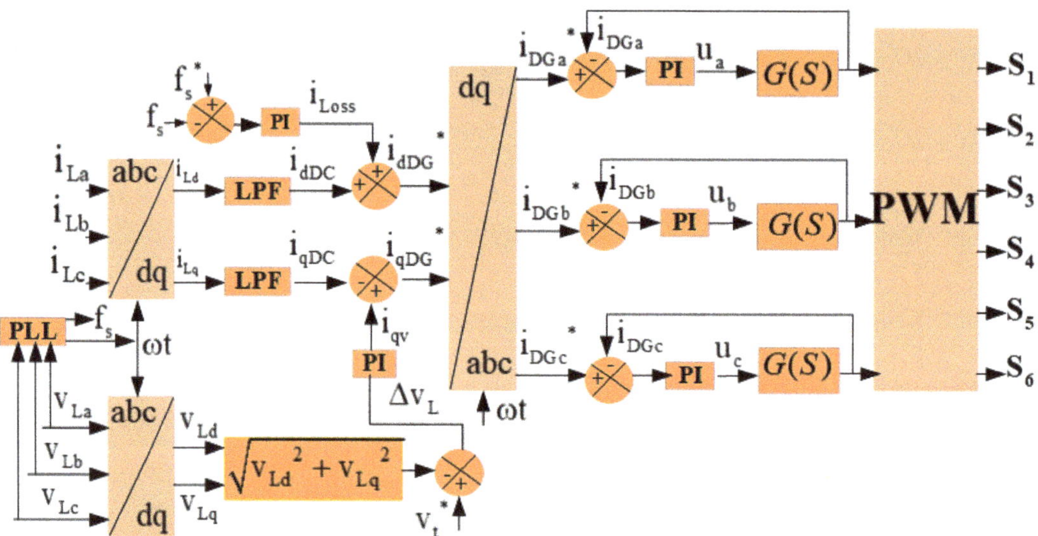

Figure 4. SRF control with virtual impedance-based active damping.

And

$$\begin{bmatrix} v_{Ld} \\ v_{Lq} \\ v_{Lo} \end{bmatrix} = \frac{2}{3} \begin{bmatrix} cos\omega t & -sin\omega t & \frac{1}{2} \\ cos\left(\omega t - \frac{2\pi}{3}\right) & -sin\left(\omega t - \frac{2\pi}{3}\right) & \frac{1}{2} \\ cos\left(\omega t + \frac{2\pi}{3}\right) & sin\left(\omega t + \frac{2\pi}{3}\right) & \frac{1}{2} \end{bmatrix} \begin{bmatrix} v_{La} \\ v_{Lb} \\ v_{Lc} \end{bmatrix} \tag{12}$$

where ωt is the phase angle and is estimated using the phase locked loop (PLL).

With help of a low pass filter (LPF), the DC components (i_{dDC}, i_{qDC}) of load current are extracted. The zero sequence i_{Lo} is taken equal to zero.

The d-axis reference DG currents ($i_{dDG}{}^*$) is estimated as

$$i_{dDG}{}^* = i_{dDC} + i_{Loss} \tag{13}$$

where i_{Loss} represents the active power component to maintain the system frequency (f_s) constant, and it compensates for the losses in the three-phase interfacing inverter. This active component is estimated as

$$i_{Loss} = K_{p5}(f_s^* - f_s) + K_{i5} \int (f_s^* - f_s) dt \tag{14}$$

where f_s^* is the system frequency reference and is equal to 60 Hz, f_s is the measured system frequency using PLL, and K_{p5} and K_{i5} represent the proportional and integral gains.

The q-axis reference DG currents ($i_{qDG}{}^*$), is estimated as

$$i_{qDG}{}^* = i_{qDC} + i_{qv} \tag{15}$$

where, i_{qv} is the reactive power component for maintaining the PCC voltage and is calculated as

$$i_{qv} = K_{p6}\Delta v_L + K_{i6} \int \Delta v_L dt \tag{16}$$

where K_{p6} and K_{i6} represent the proportional and integral gains, and (Δv_L) is calculated as

$$\Delta v_L = v_t^* - v_t$$

where v_t^* is the reference amplitude voltage at the PCC, and v_t is the measured amplitude of the PCC voltage and is calculated as

$$v_t = \sqrt{v_{Ld}{}^2 + v_{Lq}{}^2} \tag{17}$$

where v_{Ld}, v_{Lq} denote the d-q voltage of the PCC and are obtained using (12).

The reference DG currents ($i_{DGa}^*, i_{DGb}^*, i_{DGb}^*$) are obtained using the reverse Park's transformation as

$$\begin{bmatrix} i_{DGa}^* \\ i_{DGb}^* \\ i_{DGc}^* \end{bmatrix} = \begin{bmatrix} cos\omega t & sin\omega t & 1 \\ cos\left(\omega t - \frac{2\pi}{3}\right) & sin\left(\omega t - \frac{2\pi}{3}\right) & 1 \\ cos\left(\omega t + \frac{2\pi}{3}\right) & sin\left(\omega t + \frac{2\pi}{3}\right) & 1 \end{bmatrix} \begin{bmatrix} i_{dDG}^* \\ i_{qDG}^* \\ i_{oDG}^* \end{bmatrix} \tag{18}$$

The zero sequence i_{oDG}^* is taken equal to zero.

As detailed in Figure 4, the SRF control is reinforced by virtual impedance damping to solve the problem of filter resonance. Then, the equivalent transfer function of the LCL filter with dual feedback active damping is detailed in [34,35] as

$$\frac{i_{DG}(S)}{u(S)} = G(S) = \frac{1}{L_{inv}L_{DG}C_C S^3 + (k_{dc} + k_{dl})L_{DG}C_C S^2 + (L_{inv} + L_{DG})S + k_{dl}} \tag{19}$$

where k_{dc}, k_{dl} denote the damping coefficients, and L_{inv}, L_{DG} and C_C are the inductors and capacitors of the LCL filter.

The reference DG currents (i^*_{DGa}, i^*_{DGb}, i^*_{DGc}) with respective DG currents (i_{DGa}, i_{DGb}, i_{DGc}) are fed to a PWM current controller to get the switching sequences for control of the switches (S1 to S6) of the interfacing inverter.

4. Results and Discussion

To validate the effectiveness and robustness of the hybrid off-grid configuration and its composite control strategies, simulation and experimental tests are presented; they were obtained using Matlab/Simulink and hardware prototype of a 2-kW rating. Figure 5 shows the hardware prototype used for real-time validation of the complete off-grid system shown in Figure 1. It consists of (1) ABB Drive, (2) induction motor, (3) SG, (4) synchronizer, (5) transformer, (6) power converters, (7) voltage and current sensors, (8) loads, (9) protection cards, (10) dSPACE, (11) lead-acid battery pack, (12) PV emulator, (13) drive + induction motor, and (14) PMBLDCG, and (15) LCL filter.

Figure 5. Hardware prototype of hybrid wind-PV-diesel off-grid system.

4.1. Performance at the AC Side under Presence of Linear Load

Figure 6a shows the waveforms of the stator DG voltages (v_{DG}) and currents (i_{DG}), load voltages (v_L), and currents (i_L), and Figure 6b shows the zoom of the waveforms shown in Figure 6a between t = 1.1 s and t = 1.3 s. In this test, a linear load is connected at the PCC. One observes in Figure 6a,b that the stator voltages of the SG are well regulated and are equal to the reference 208 V. The voltage and current at the PCC are constant and sinusoidal. The system frequency is well regulated and is equal to 60 Hz.

4.2. Performance at the DC Side at Solar Irradiation and Wind Speed Change

In Figure 7, the waveforms of the common DC-link voltage (V_{dc}) and its reference (V^*_{dc}); ESS current (i_b) and its reference i^*_b; the state of charge of the ESS (SoC%); the output PV current (i_{PV}); the stator voltage of the phase 'a' of the PMBLDCG (V_{sa}) and the current (i_{sa}); and the DC-WT current (i_{dc}), which is measured at the output of the three-phase diode bridge, are presented during dynamics of RESs generation.

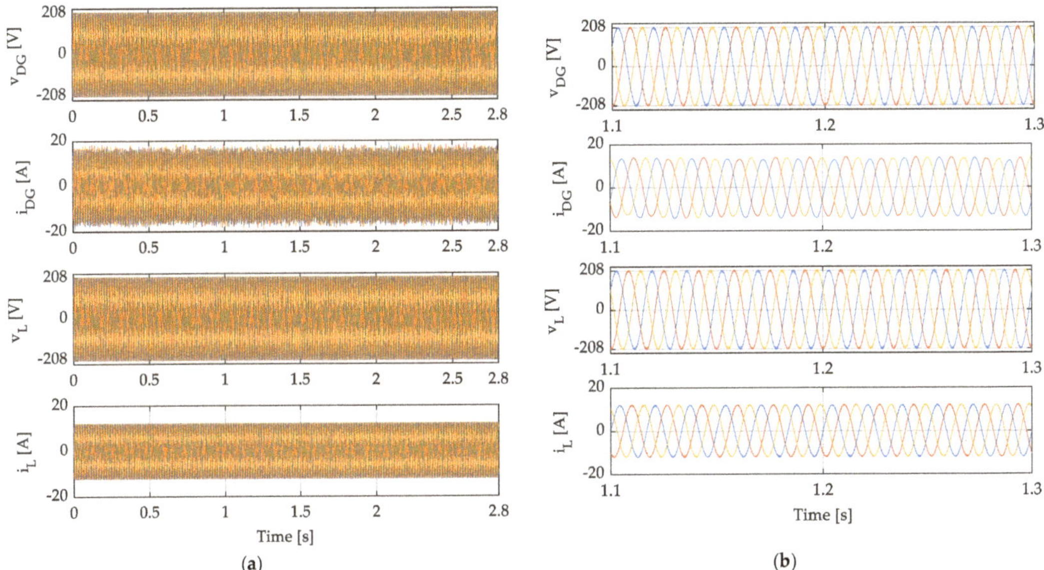

Figure 6. Performance under fixed linear load at the AC side and (**b**) Zoom of (**a**) between t = 1.1 s to 1.3 s.

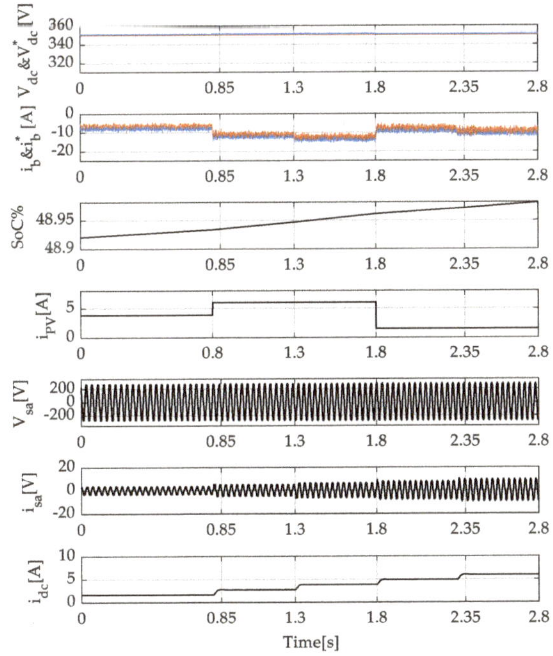

Figure 7. Performance at the DC side under wind and solar irradiation change of the common DC-link voltage (V_{dc}) and its reference (V_{dc}^*); ESS current (i_b) and its reference i_b^*; the state of charge of the ESS (SoC%); the output PV current (i_{PV}); the stator voltage of the phase 'a' of the PMBLDCG (V_{sa}) and the current (i_{sa}); and the DC-WT current (i_{dc}).

One observes that the common DC-link voltage is well regulated and is following its reference, which is equal to 350 V and is not much affected when the system is subjected to sudden solar irradiation change at t = 0.8 s and t = 1.8 s, and when the wind speed suddenly increases at t = 0.85 s, t = 1.3 s, t = 1.8s, and t = 2.35 s, respectively. One observes that ESS currents follow its reference during weather condition changes. Seeing that the SoC% of ESS is less than 50% as detailed in Table 1 (operating mode 2), DG provides power to the connected linear load as shown in Figure 6 and charges the ESS. In addition, all generated power from the WT and PV panel is used to charge the ESS, which is why the ESS current varies with the variation of the WT and PV panel currents. One observes clearly that one extracts the MPP from the PV panel without using any power converter and from the WT using only a three-phase-diode bridge and boost converter. It is demonstrated that we can operate the WT efficiently under variable speed conditions and provide accurate and fast convergence to the maximum global operating point without speed sensors by applying the iterative interpolation method with perturbation and observation (P&O) technique. Based on the obtained results, one may conclude that the developed control for the DC-DC buck-boost converter shown in Figure 3, and the one for the DC-DC boost converter shown in Figure 2, operate well under all conditions without any saturation issue and with high efficiency.

4.3. Generated and Consumed Active and Reactive Powers

In Figure 8, the waveforms of active and reactive powers of DG (P_{DG}, Q_{DG}), load (P_L, Q_L), three-phase interfacing inverter (P_{inv}, Q_{inv}), generated active power from WT (P_W), and PV panel (P_{PV}), and from the ESS (P_{ESS}), are demonstrated. One observes that the balance in power in the hybrid off-grid system is perfectly achieved, and all generated power in this operating mode (operating mode 2) is provided to the load and to charge the ESS because the SoC% is less than 50%. So, on the AC side, the DG is supplying the load directly and the difference of power is provided to the ESS through the three-phase interfacing inverter, which is why the active and reactive power is with a negative sign. On the DC side, all generated power from the WT and the PV array is used to charge the ESS, which is why the P_{ESS} is with a negative sign; it is increasing and decreasing based on the variation of the wind speed and solar irradiation. It is observed that the power balance is achieved without any issue, which confirms the robustness of the indirect control developed for the DC-DC buck-boost converter based on the double loop strategy.

4.4. Performance at PCC under the Presence of Nonlinear Loads

In Figures 9a and 10a, the waveforms of the common DC voltage (V_{dc}) and its reference (V_{dc}^{*}), currents of the phase 'a' of DG (i_{DGa}), load (i_{La}), and inverter (i_{inva}) are demonstrated, and in Figure 9b,c and Figure 10b,c, the harmonic spectrum of load and DG currents are presented. One observes that the interfacing inverter acts as a power active filter under the presence of RL and RC both types of nonlinear loads. It compensates for harmonics and balances the DG current. One observes that the common DC-link voltage is maintained constant and follows its reference, which is equal to 350 V. One sees clearly from the spectra of harmonics shown in Figures 9b and 10b the presence of the 5th and 7th order of harmonics in the load current. The total harmonics distortion (THD) of the load current is equal to 26.9% in Figure 9b for RC type nonlinear load and 26.05% in Figure 10b for RL type nonlinear load. The THD of the DG current as demonstrated in Figures 9c and 10c is equal to 2.31% and 1.88%, respectively. This achieves the limit of IEEE Std 519-1992. One observes that the 5th and 7th order harmonics are mitigated at the level of DG current in the presence of both nonlinear loads. This proves that the stator of the SG is protected against 5th and 7th order harmonics under the presence of all nonlinear loads, which confirms the robustness of the SRF control with virtual impedance-based active damping, which is developed for the three-phase interfacing inverter.

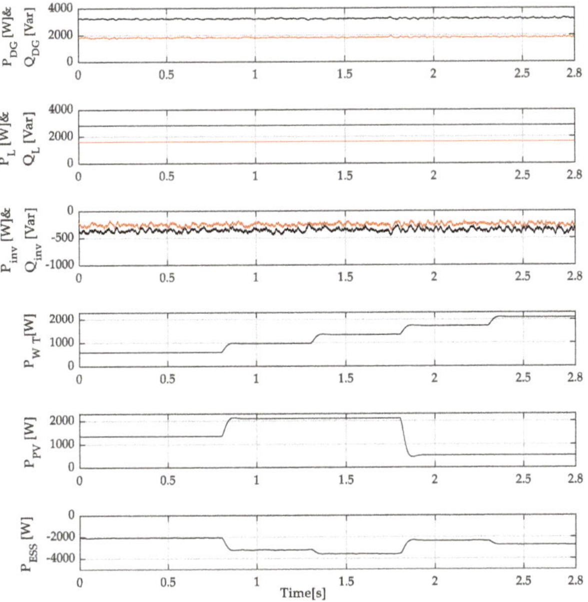

Figure 8. Global power demand and generation in the hybrid wind-PV-diesel off-grid system; theactive and reactive powers of DG (P_{DG}, Q_{DG}), load (P_L, Q_L), three-phase interfacing inverter (P_{inv}, Q_{inv}), generated active power from WT (P_W), and PV panel (P_{PV}), and from the ESS (P_{ESS}).

4.5. Experimental Results of the DG under the Presence of Linear Load

Figure 11a–c show the performance of the DG under the presence of linear load. One observes that the terminal stator voltages (v_{DGab} and v_{DGbc}) are sinusoidal and regulated at their rated value, which equals 208 V at the primary of the transformer (element no. 5 in Figure 5 and 50 V at the secondary as shown in Figure 11c. The connected load is supplied with a constant current at a fixed frequency (i.e., 60 Hz). The DC current applied to the rotor winding of SG is constant and is equal to 0.8 A.

4.6. Performance under Load Variation and Presence of Linear Load

In Figure 12a–c, the waveforms of the common DC-link voltage (v_{dc}), output DG current of phase 'a' (i_{DGa}), the generated current from the PV panel and WT ($i_{WT} + i_{PV}$), load current of phase 'a' (i_{La}), ESS current (i_b), and line PCC voltage (v_{Lab}) are presented. It is observed in Figure 12a that the load is suddenly connected to the system at t = 0.72 ms, and the DG stabilizes only after t = 0.12 s, and in this period, the ESS provides the difference. One observes that the common DC-link voltage is well regulated, and it is not affected during the transition, which confirms the robustness of the outer control loop of the DC-DC buck-boost converter. It is observed in Figure 12b, that ESS provides the difference of power between t = 0 s to t = 0.2 s, and the current becomes equal to zero when the load is disconnected at t = 0.2 s. In Figure 12c, the ESS current is increased more when the load is suddenly increased at t = 0.2 s. In all the cases, the load is supplied without interruption, and the power is balanced in the hybrid off-grid system. The common DC-link voltage is well regulated without deviation (overshoot and undershoot), which confirms the robustness of indirect control developed for the DC-DC buck-boost converter.

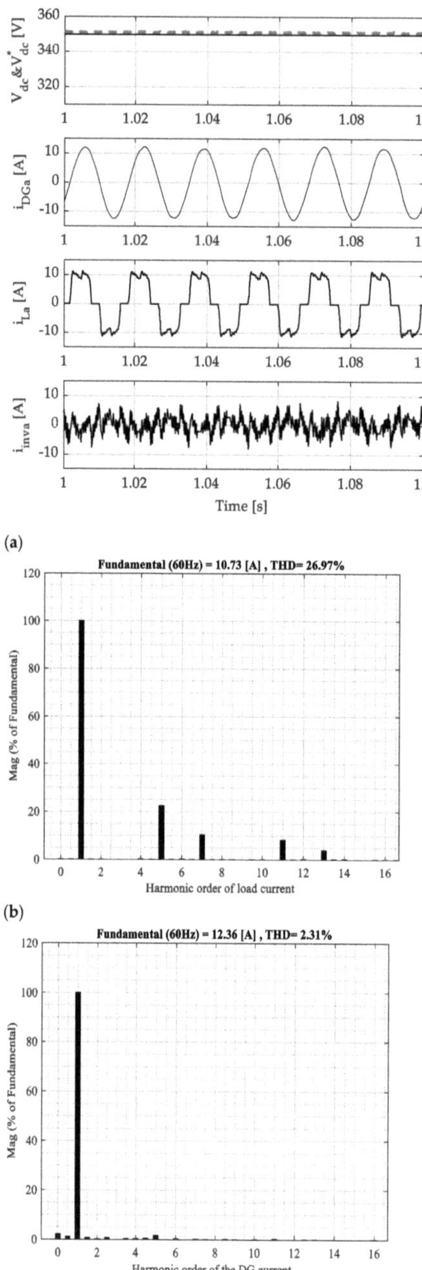

(a)

(b)

(c)

Figure 9. (a) Waveforms of the DG, load, and inverter currents under the presence of nonlinear load type RC; (b) harmonic spectrum of load current; and (c) harmonic spectrum of DG current.

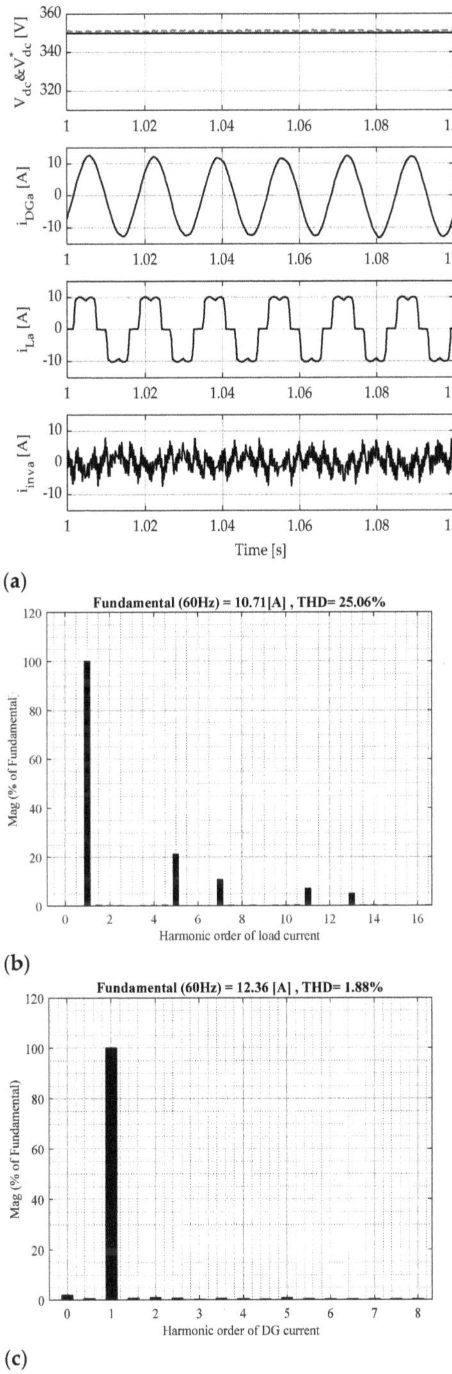

Figure 10. (**a**) Waveforms of DG, load, and inverter currents under the presence of nonlinear load type RL; (**b**) harmonic spectrum of load current; and (**c**) harmonic spectrum of DG current.

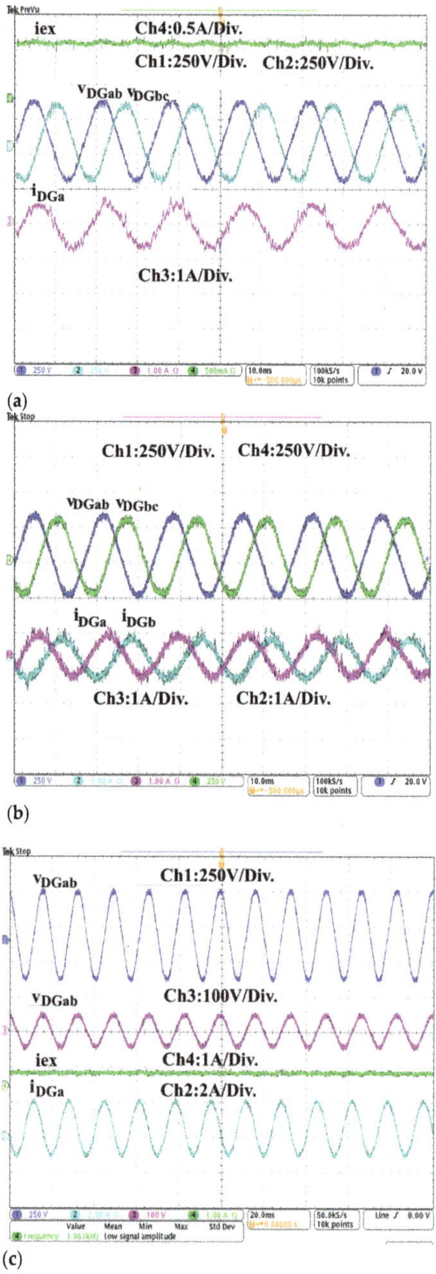

Figure 11. Steady-state performance of (**a**) excitation current (i_{ex}), SG terminal stator voltages (v_{DGab}, v_{DGbc}) at the primary of the transformer and stator current (i_{DGa}) of phase 'a'; (**b**) SG terminal stator voltages (v_{DGab}, v_{DGbc}) at primary of transformer and stator currents (i_{DGa}, i_{DGb}) of phase 'a' and 'b'; and (**c**) SG terminal stator voltage (v_{DGab}) at primary of the transformer and at secondary (v_{DGab}), excitation current (i_{ex}), and stator current (i_{DGa}).

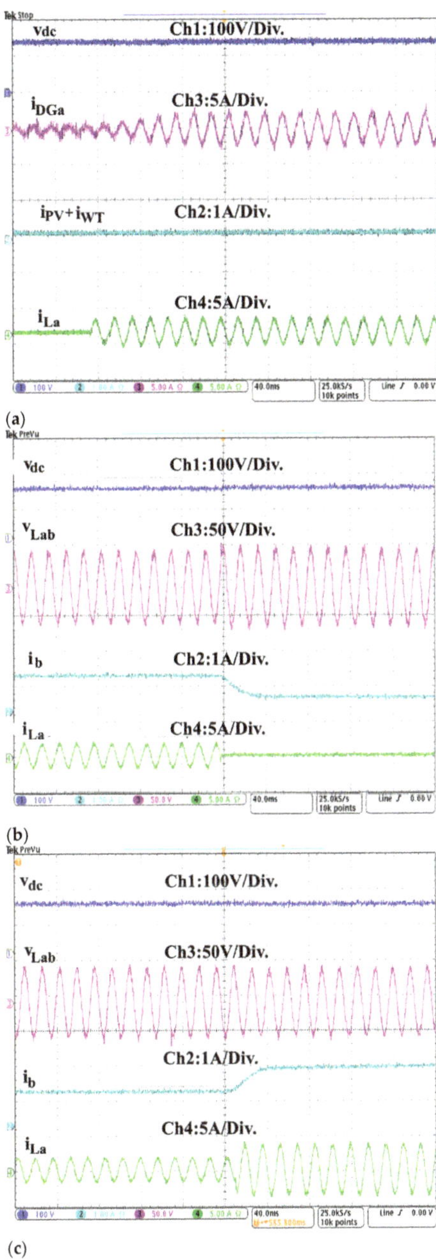

Figure 12. Dynamic performance of system: (**a**) common DC-link voltage (v_{dc}), DG current (i_{DGa}) of phase 'a', sum of PV current and DC-WT current ($i_{PV} + i_{WT}$), and load current (i_{La}) of the phase 'a' at sudden switching ON of linear load at t = 0.08 s; (**b**) common DC-link voltage (v_{dc}), load voltage of phase 'a' (v_{Lab}), ESS current (i_b), and load current of phase 'a' (i_{La}) under sudden switching off of linear load at t = 0.2 s; and (**c**) common DC-link voltage (v_{dc}), load voltage of phase 'a'(v_{Lab}), ESS current (i_b), and load current of phase 'a' (i_{La}) under sudden increasing of linear load at t = 0.2 s.

4.7. Performance in Presence of RC and RL Types Nonlinear Loads

In Figures 13a and 14a, the common waveforms of DC-link voltage (v_{dc}), load current of phase (i_{La}), DG current (i_{DGa}), and the output inverter current (i_{inva}) of phase 'a' in the presence of nonlinear loads type RL and type RC are presented. In Figure 13b,c and Figure 14b,c, the harmonics spectra of the load and DG currents are demonstrated. One observes clearly in Figures 13a and 14a that the DG currents are sinusoidal and balanced in presence of the both type of nonlinear loads, and the common DC-link voltage is well regulated at its set value, which is equal to 120 V. It is observed that the inverter operates as an active filter; it mitigates harmonics and balances the DG current. This confirms the robustness of the proposed indirect control strategy for the DC-DC buck-boost converter and the SRF with a virtual impedance-based-active damping control strategy for the three-phase interfacing inverter. As demonstrated in Figures 13c and 14c, the 5th and the 7th order harmonics are mitigated and THD of DG currents is less than 5% in the presence of both types of nonlinear loads, which confirms that the SG of the DG is protected against 5th and 7th order harmonics and proves the robustness of the developed control strategy for the three-phase interfacing inverter.

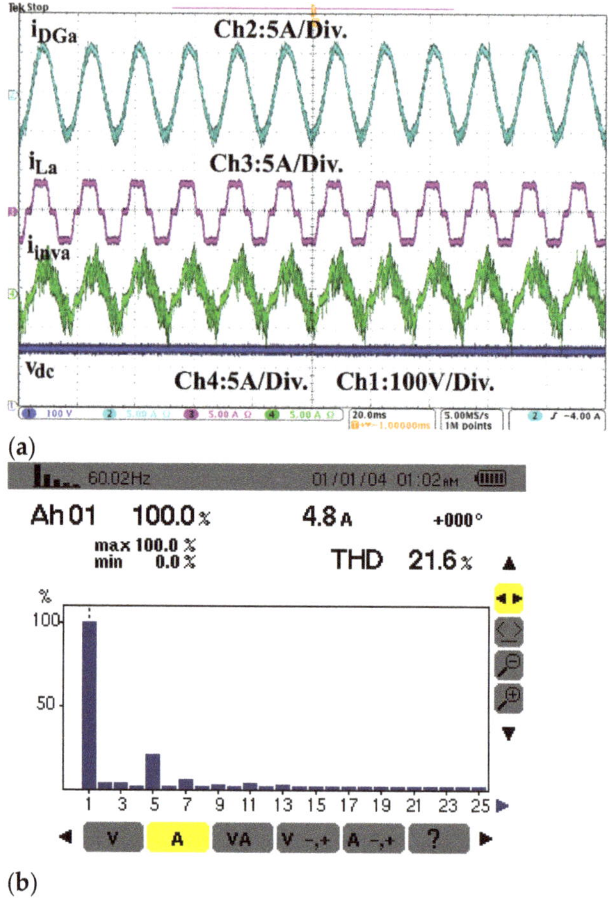

Figure 13. *Cont.*

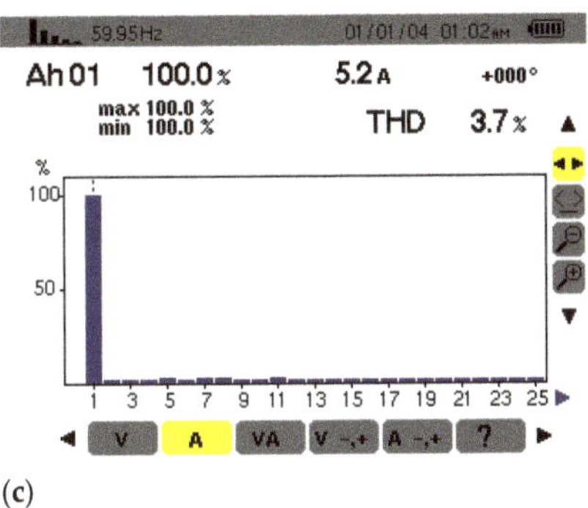

(c)

Figure 13. (**a**) Performance in the presence of nonlinear load (RL type), (**b**) harmonic spectrum of load current, and (**c**) harmonic spectrum of DG current.

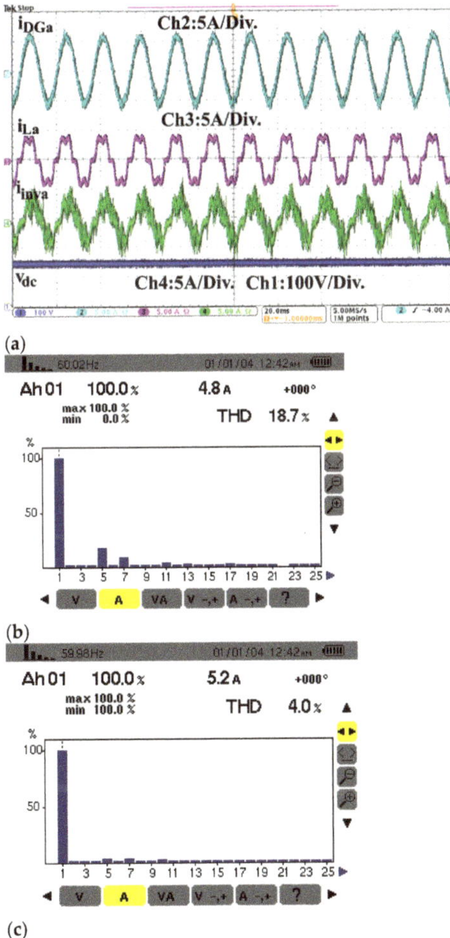

(a)

(b)

(c)

Figure 14. (**a**) Performance in the presence of nonlinear load (RC type), (**b**) harmonic spectrum of load current, and (**c**) harmonic spectrum of DG current.

4.8. Performance of the WT at Wind Speed Change

The waveforms of the line PCC voltage (v_{Lab}), load current (i_{La}), output WT current at the DC side, and ESS current (i_b) are presented in Figure 15. This test is realized by maintaining a constant load and applying different speeds to the PMBLDCG. One observes clearly that the output DC current, which is measured at the DC side of the WT, varies with the variation of the wind speed. Seeing that the DG is running at this time, the difference of generated power is taken by ESS. The ESS current is increased at t = 1.8 s and increased further at t = 3 s, t = 6 s, and at t = 12.2 s. One sees clearly that perturbation and observation (P&O) with variable steps of DC voltage based on the secant method performs well during wind variation, and all transitions are realized without any overshoot in current or voltage. It is observed that the PCC voltage is well regulated, and load is continually supplied without any interruption during the transition and during rotor speed variation. This proves the effective operation of the MPPT technique without sensing the rotor speed proposed for the variable speed wind turbine based on PMBLDCG.

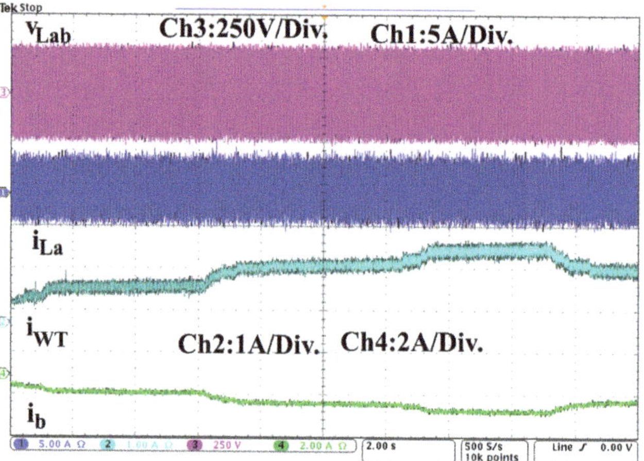

Figure 15. Performance at the variation of wind speed at fixed linear load.

In Figure 16a, the waveforms of stator voltage (v_{sa}) of phase 'a', WT output DC voltage (v_{WT}) and current (i_{WT}), and stator current (i_{sa}) of phase 'a' are presented. In Figure 16, (b) the zoomed waveforms of (a) between t = 2.64 s and t = 2.88 s are shown. One observes that experimental results are similar to the simulation results where the output DC voltage of the WT varies slightly with the variation of the wind speed. It is observed that the output stator current and the current measured at the output of three-phase diode-bridge increase at t = 0.8 s and increase more at t = 2.8 s with the increasing of the rotor speed. This validates the desired operation of the MPPT technique without mechanical speed measurement.

4.9. Performance of the PV Panel at Solar Irradiation Change

The waveforms of the PCC voltage (v_{DGa}) of the phase 'a', the PV current (i_{PV}), ESS current (i_b), and the common DC link voltage are shown in Figure 17. This test is realized under solar irradiation change to test the performance of the proposed approach to achieve high efficiency from PV array without using any MPPT technique. One observes that the common DC-link voltage is regulated constant at its rated value, which is equal to the sum of the output voltage of the PVs connected in series. The system is subjected to solar irradiation change at t = 0.4 s, t = 0.72 s, and 1.16 s. One observes that by maintaining the common DC link voltage constant, one can easily extract the maximum of current from the PV panel during solar irradiation change. It is observed that the ESS current increases with the increase of the PV panel current, and V_{dc} is maintained constant; this proves the desired operation of the outer and inner control loops of the indirect control, which are designed to achieve MPPT from PV and balance the power in the off-grid system.

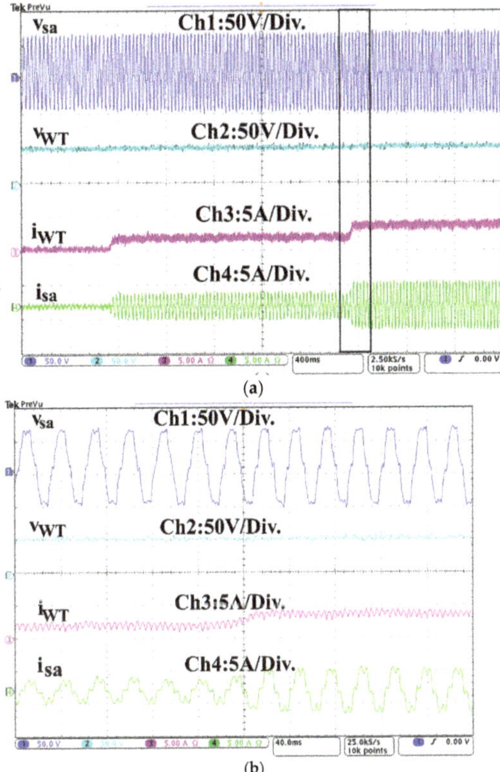

Figure 16. Waveforms of WT side for: (**a**) stator voltage (v_{sa}) of phase 'a', output DC voltage (v_{WT}) and current (i_{WT}), and stator current (i_{sa}) and (**b**) its zoomed waveform.

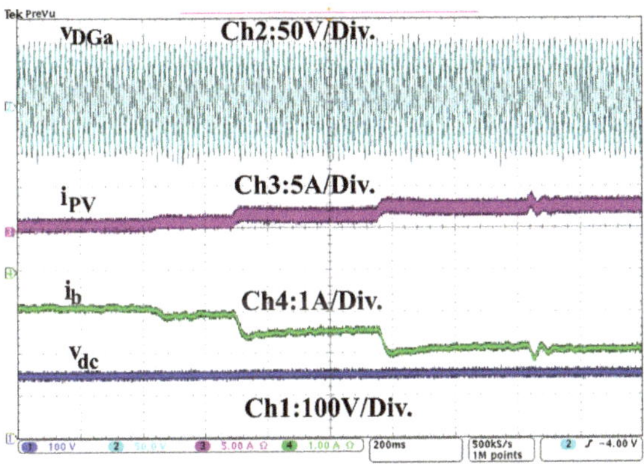

Figure 17. Waveforms of PCC voltage (v_{DGa}) at secondary of the transformer of output PV (i_{PV}), ESS current (i_b), and common DC link voltage (v_{dc}).

5. Conclusions

The performance of the developed composite control strategy with reduced sensors for a PV-wind-diesel-based off-grid power generation system has been presented in this research work. The developed off-grid configuration has reduced the number of power converters to make it an effective, low-cost option. To achieve stable operation under disturbance and balance the power in the system, the developed control strategy for the DC-DC buck-boost converter is reinforced by the inner control loop. Furthermore, for better attenuation of switching harmonics without additional losses, active damping solution is developed and implemented. The obtained simulation and experimental results show satisfactory performance under the change in solar irradiation and wind speed, as well as balanced linear and nonlinear loads. It has been demonstrated that the perturbation and observation (P&O) technique with variable steps of DC voltage based on the secant method performs well during sudden wind speed variations without any saturation issue and use of mechanical speed sensor. The extraction of maximum power from PV panels without the MPPT algorithm has also been demonstrated. Furthermore, the control for a three-phase interfacing inverter using SRF control with virtual impedance-based active damping has been demonstrated to improve the power quality at PCC and power flow management in the system. In addition, the 5th and 7th order-harmonics are mitigated, and DG is operated perfectly without any challenge under the presence of nonlinear loads. Therefore, it is concluded that the developed composite control strategy for PV-wind-diesel-based off-grid power generation system performs well in the presence of severe real-time conditions. Up to now, the performance of the PV array has been tested under a fixed temperature of 25 °C and variable solar irradiation; therefore, the system efficiency under different temperatures needs evaluation. Furthermore, the high-frequency voltage stress on the DC side and the deterioration of the frequency and voltage stability under power imbalances need further evaluation.

Author Contributions: M.R. proposed the original idea and carried out the main research tasks, simulation, implementation, and writing. F.D. conducted experiments, and M.I. designed system elements. H.I. carried out formal analysis and reviews, A.C. and B.S. supervised and carried out reviews; S.S. performed the design of system element, writing, review, and editing. All authors have read and agreed to the published version of the manuscript.

Funding: This research received no external funding.

Institutional Review Board Statement: Not applicable.

Informed Consent Statement: Not applicable.

Data Availability Statement: Not applicable.

Acknowledgments: The authors wish to thank RQEI (Réseau québécois sur l'énergie intelligente) for their support.

Conflicts of Interest: The authors declare no conflict of interest.

Define Abbreviations

Symbol	Description
WT	Wind turbine
PV	Photovoltaic array
ESS	Energy storage system
DG	Diesel generator
PMBLDCG	Permanent magnet brushless DC generator
i_{sa}, i_{sb}, and i_{sc}	Stator currents of the PMBLDCG
C_{WT}	The capacitor at the output of the diode bridge
V_{WT}	DC voltage of the WT

L_{WT}	The inductor of the DC-DC boost converter for WT side
i_{WT}	DC current of the WT
V_{PV}	Output PV voltage
i_{PV}	Output PV current
v_b	Battery voltage
L_b	Inductor that connects the battery to the DC-DC buck-boost converter
V_{dc}	Common DC-link voltage
V_{inva}, V_{invb}, and V_{invc}	Output voltages of the interfacing inverter
i_{inva}, i_{invb}, and i_{invc}	Output currents of the interfacing inverter
V_{ca}, V_{cb}, and V_{cc}	Voltages of the output filter
R_c and C_C	Resistance and capacitor of the output filter
i_{cc}	The current of the output filter
L_{inv} and L_{DG}	Inductors of the output filter
V_{La}, V_{Lb}, and V_{Lc}	Load voltages
i_{La}, i_{Lb}, and i_{Lc}	Load currents
i_{DGa}, i_{DGb}, and i_{DGc}	Diesel generator currents
AVR	Automatic voltage regulator
DE	Diesel engine
P&O	Perturbation and observation technique
V_{WTmax}	The maximum voltage obtained using the P&O technique
ΔV_{WT}	Step of V_{WT} variation
$V_{WT}{}^*$	Reference DC voltage of the WT
ΔV	WT DC voltage error value
PI	Proportional integral regulator
$i_{WT}{}^*$	Reference DC current of the WT
Δi_{WT}	WT DC current error value
d_{WT}	Control signal
S_1 to S_9	Power electronic switches *(insulated-gate bipolar transistor (IGBT))* of the power converters
PWM	Pulse-width modulation
$V_{dc}{}^*$	Reference of the common DC-link voltage
ΔV_{dc}	Error value of the common DC-link voltage
$i_b{}^*$	Reference battery current
Δi_b	Error value of battery current
d_b	Control signal
f_s	System frequency
$f_s{}^*$	Reference of the system frequency
l_{Loss}	losses of active power
d-q	Direct and quadrature axis
LPF	Low pass filter
PLL	Phased locked loop
ωt	Angular frequency
$i_{DGa}{}^*$, $i_{DGb}{}^*$ and $i_{DGc}{}^*$	Reference of DG currents
$G(s)$	Transfer function of LCL filter

References

1. Hidalgo-Leon, R.; Amoroso, F.; Litardo, J.; Urquizo, J.; Torres, M.; Singh, P.; Soriano, G. Impact of the Reduction of Diesel Fuel Subsidy in the Design of an Off-Grid Hybrid Power System: A Case Study of the Bellavista Community in Ecuador. *Energies* **2021**, *14*, 1730. [CrossRef]
2. Mokhtara, C.; Negrou, B.; Settou, N.; Settou, B.; Samy, M.M. Design optimization of off-grid Hybrid Renewable Energy Systems considering the effects of building energy performance and climate change: Case study of Algeria. *Energy* **2021**, *219*, 119605. [CrossRef]
3. Dubuisson, F.; Rezkallah, M.; Ibrahim, H.; Chandra, A. Real-Time Implementation of the Predictive-Based Control with Bacterial Foraging Optimization Technique for Power Management in Standalone Microgrid Application. *Energies* **2021**, *14*, 1723. [CrossRef]

4.   Rezkallah, M.; Singh, S.; Chandra, A.; Singh, B.; Ibrahim, H. Off-Grid System Configurations for Coordinated Control of Renewable Energy Sources. *Energies* **2020**, *13*, 4950. [CrossRef]
5.   Rezkallah, M.; Chandra, A.; Singh, B.; Singh, S. Microgrid: Configurations, Control and Applications. *IEEE Trans. Smart Grid* **2017**, *10*, 1290–1302. [CrossRef]
6.   Sawle, Y.; Gupta, S.; Bohre, A.K. Review of hybrid renewable energy systems with comparative analysis of off-grid hybrid system. *Renew. Sustain. Energy Rev.* **2018**, *81*, 2217–2235. [CrossRef]
7.   Upadhyay, S.; Sharma, M. A review on configurations, control and sizing methodologies of hybrid energy systems. *Renew. Sustain. Energy Rev.* **2014**, *38*, 47–63. [CrossRef]
8.   Nehrir, M.H.; Wang, C.; Strunz, K.; Aki, H.; Ramakumar, R.; Bing, J.; Miao, Z.; Salameh, Z. A Review of Hybrid Renewable/Alternative Energy Systems for Electric Power Generation: Configurations, Control, and Applications. *IEEE Trans. Sustain. Energy* **2011**, *2*, 392–403. [CrossRef]
9.   Beleiu, H.G.; Maier, V.; Pavel, S.G.; Birou, I.; Pică, C.S.; Dărab, P.C. Harmonics Consequences on Drive Systems with Induction Motor. *Appl. Sci.* **2020**, *10*, 1528. [CrossRef]
10. Rezkallah, M.; Chandra, A.; Ibrahim, H.; Singh, B. Implementation of Two-Level Control Coordinate for Seamless Transfer in Standalone Microgrid. In Proceedings of the 2019 IEEE Industry Applications Society Annual Meeting, Baltimore, MD, USA, 29 September–3 October 2019; Volume 75, pp. 1–6. [CrossRef]
11. Park, B.; Jaehyeong, L.; Hangkyu, Y.; Gilsoo, J. Harmonic Mitigation Using Passive Harmonic Filters: Case Study in a Steel Mill Power System. *Energies* **2021**, *14*, 2278. [CrossRef]
12. Hoon, Y.; Mohd, A.M.R.; Muhammad, A.A.M.Z.; Mohamad, A.M.Z. Shunt active power filter: A review on phase synchronization control techniques. *Electronics* **2019**, *8*, 7. [CrossRef]
13. Bielecka, A.; Wojciechowski, D. Stability Analysis of Shunt Active Power Filter with Predictive Closed-Loop Control of Supply Current. *Energies* **2021**, *14*, 2208. [CrossRef]
14. Bhim, S.; Chandra, A.; Kamal, A.L.-H. *Power Quality: Problems and Mitigation Techniques*; John Wiley & Sons: Hoboken, NJ, USA, 2014.
15. Mehmood, S.; Qureshi, A.; Kristensen, A.S. Risk Mitigation of Poor Power Quality Issues of Standalone Wind Turbines: An Efficacy Study of Synchronous Reference Frame (SRF) Control. *Energies* **2020**, *13*, 4485. [CrossRef]
16. Singh, B.; Sharma, S. SRF theory for voltage and frequency control of IAG based wind power generation. In Proceedings of the International Conference on Power Systems, Kharagpur, India, 27–29 December 2009; IEEE: Piscataway, NJ, USA, 2009; pp. 1–6.
17. Naderipour, A.; Asuhaimi Mohd Zin, A.; Bin Habibuddin, M.H.; Miveh, M.R.; Guerrero, J.M. An improved synchronous reference frame current control strategy for a photovoltaic grid-connected inverter under unbalanced and nonlinear load conditions. *PLoS ONE* **2017**, *12*, e0164856. [CrossRef] [PubMed]
18. Benhalima, S.; Miloud, R.; Chandra, A. Real-Time Implementation of Robust Control Strategies Based on Sliding Mode Control for Standalone Microgrids Supplying Non-Linear Loads. *Energies* **2018**, *11*, 2590. [CrossRef]
19. Lopez-Santos, O.; Urrego-Aponte, J.O.; Lezama, S.T.; Almansa-López, J.D. Control of the Bidirectional Buck-Boost Converter Operating in Boundary Conduction Mode to Provide Hold-Up Time Extension. *Energies* **2018**, *11*, 2560. [CrossRef]
20. Ramos-Paja, C.A.; Bastidas-Rodríguez, J.D.; González, D.; Acevedo, S.; Peláez-Restrepo, J. Design and Control of a Buck–Boost Charger-Discharger for DC-Bus Regulation in Microgrids. *Energies* **2017**, *10*, 1847. [CrossRef]
21. Yu, S.Y.; Kim, H.J.; Kim, J.H.; Han, B.M. SoC-based output voltage control for BESS with a lithium-ion battery in a stand-alone DC microgrid. *Energies* **2016**, *11*, 924. [CrossRef]
22. Zhu, Y.; Kim, M.K.; Wen, H. Simulation and Analysis of Perturbation and Observation-Based Self-Adaptable Step Size Maximum Power Point Tracking Strategy with Low Power Loss for Photovoltaics. *Energies* **2018**, *12*, 92. [CrossRef]
23. Rezkallah, M.; Hamadi, A.; Chandra, A.; Singh, B. Design and Implementation of Active Power Control With Improved P&O Method for Wind-PV-Battery-Based Standalone Generation System. *IEEE Trans. Ind. Electron.* **2017**, *65*, 5590–5600. [CrossRef]
24. Jung, Y.; So, J.; Yu, G.; Choi, J. Improved perturbation and observation method (IP&O) of MPPT control for photovoltaic power systems. In Proceedings of the Conference Record of the Thirty-First IEEE Photovoltaic Specialists Conference, Lake Buena Vista, FL, USA, 3–7 January 2005; pp. 1788–1791.
25. Gunasekaran, M.; Krishnasamy, V.; Selvam, S.; Almakhles, D.J.; Anglani, N. An Adaptive resistance perturbation based MPPT algorithm for Photovoltaic applications. *IEEE Access* **2020**, *8*, 1. [CrossRef]
26. Pathak, G.; Singh, B.; Panigrahi, B.K. Isolated microgrid employing PMBLDCG for wind power generation and synchronous reluctance generator for DG system. In Proceedings of the 2014 IEEE 6th India International Conference on Power Electronics (IICPE), Kurukshetra, India, 8–10 December 2014; pp. 1–6.
27. Sharma, R.; Singh, B. MDSOGI-FLL Control for SyRG-PMBLDCG-BES-PV Based Microgrid. In Proceedings of the IEEE Industrial and Commercial Power Systems Europe, Genova, Italy, 11–14 June 2019; pp. 1–6.
28. Pathak, G.; Singh, B.; Panigrahi, B.K. Control of Wind-Diesel Microgrid Using Affine Projection-Like Algorithm. *IEEE Trans. Ind. Inform.* **2016**, *12*, 524–531. [CrossRef]
29. Chen, Y.-M.Y.; Liu, C.S.; Hung, C.; Cheng, C.-S. Multi-input inverter for grid-connected hybrid PV/wind power system. *IEEE Trans. Power Electron.* **2007**, *2*, 1070–1077. [CrossRef]
30. Chun, S.; Kwasinski, A. Analysis of classical root-finding methods applied to digital maximum power point tracking for sustainable photovoltaic energy generation. *IEEE Trans. Power Electron.* **2011**, *26*, 3730–3743. [CrossRef]

31. Hosseini, S.H.; Farakhor, A.; Haghighian, S.K. Novel algorithm of maximum power point tracking (MPPT) for variable speed PMSG wind generation systems through model predictive control. In Proceedings of the 2013 8th International Conference on Electrical and Electronics Engineering (ELECO), Bursa, Turkey, 28–30 November 2013; pp. 243–247.
32. Rezkallah, M.; Chandra, A.; Saad, M.; Tremblay, M.; Singh, B.; Singh, S.; Ibrahim, H. Composite Control Strategy for a PV-Wind-Diesel based Off-Grid Power Generation System Supplying Unbalanced Non-Linear Loads. In Proceedings of the 2018 IEEE Industry Applications Society Annual Meeting (IAS), Portland, OR, USA, 23–27 September 2018; pp. 1–6.
33. Rezkallah, M.; Chandra, A.; Saad, M.; Tremblay, M.; Singh, B.; Singh, S.; Ibrahim, H. Design and Implementation of Decentralized Control for Distributed generation based Off-grid System. In Proceedings of the 2020 IEEE International Conference on Power Electronics, Smart Grid and Renewable Energy (PESGRE2020), Cochin, India, 2–4 January 2020; pp. 1–6.
34. Bao, C.; Ruan, X.; Wang, X.; Li, W.; Pan, D.; Weng, K. Step-by-Step Controller Design for LCL-Type Grid-Connected Inverter with Capacitor–Current-Feedback Active-Damping. *IEEE Trans. Power Electron.* **2014**, *29*, 1239–1253. [CrossRef]
35. Liu, F.; Zhou, Y.; Duan, S.; Yin, J.; Liu, B.; Liu, F. Parameter Design of a Two-Current-Loop Controller Used in a Grid-Connected Inverter System with LCL Filter. *IEEE Trans. Ind. Electron.* **2009**, *56*, 4483–4491. [CrossRef]

clean technologies

Article

Interphase Power Flow Control via Single-Phase Elements in Distribution Systems

Piyapath Siratarnsophon [1], Vinicius C. Cunha [2], Nicholas G. Barry [1] and Surya Santoso [1,*]

[1] Department of Electrical and Computer Engineering, University of Texas at Austin, Austin, TX 78712, USA; piyapath@utexas.edu (P.S.); nicholas.barry@utexas.edu (N.G.B.)

[2] Department of Systems and Energy, University of Campinas, Campinas 13083-970, Brazil; vcunha@dsee.fee.unicamp.br

* Correspondence: ssantoso@mail.utexas.edu

Abstract: The capability of routing power from one phase to another, interphase power flow (IPPF) control, has the potential to improve power systems efficiency, stability, and operation. To date, existing works on IPPF control focus on unbalanced compensation using three-phase devices. An IPPF model is proposed for capturing the general power flow caused by single-phase elements. The model reveals that the presence of a power quantity in line-to-line single-phase elements causes an IPPF of the opposite quantity; line-to-line reactive power consumption causes real power flow from leading to lagging phase while real power consumption causes reactive power flow from lagging to leading phase. Based on the model, the IPPF control is proposed for line-to-line single-phase power electronic interfaces and static var compensators (SVCs). In addition, the control is also applicable for the line-to-neutral single-phase elements connected at the wye side of delta-wye transformers. Two simulations on a multimicrogrid system and a utility feeder are provided for verification and demonstration. The application of IPPF control allows single-phase elements to route active power between phases, improving system operation and flexibility. A simple IPPF control for active power balancing at the feeder head shows reductions in both voltage unbalances and system losses.

Keywords: distribution system; microgrids; power quality; power system management; power system reliability; smart grids

Citation: Siratarnsophon, P.; Cunha, V.C.; Barry, N.G.; Santoso, S. Interphase Power Flow Control via Single-Phase Elements in Distribution Systems. *Clean Technol.* **2021**, *3*, 37–58. https://doi.org/10.3390/cleantechnol3010003

Received: 3 December 2020
Accepted: 11 January 2021
Published: 13 January 2021

1. Introduction

AC power systems employ three-phase power technologies for economic reasons. Even though power in each phase is naturally independent, i.e., loads are supplied by the generation of the same phase, the capability for routing power between phases or interphase power flow (IPPF) control, can improve flexibility and operation for distribution systems. Power can be routed from heavily to lightly loaded phases for load in-balance compensation. As a result, system losses can be reduced [1,2] while improving utilization of power equipment. IPPF control capability also improves the system operation, especially in microgrids. In critical scenarios, a phase may experience load-generation imbalance due to line trips or insufficient generation. With IPPF control capability, the power of the interrupted phase can be routed from the other phases to maintain load-generation balance and system stability.

Among others, instantaneous symmetrical components [3], current physical components [4], and the power unbalance compensation via static var compensators (SVCs) in [5] are the leading theories related to power routing. However, they only focus on load balancing and are not applicable for general power routing applications. Moreover, these theories are developed for three-phase devices and not for single-phase devices.

From the hardware development perspective, line switches and three-phase flexible alternating current transmission systems (FACTs) are the devices currently considered for interphase power routing control applications. Line switches and tie-lines can be utilized to swap a part of a heavily loaded phase with a part of a lightly loaded phase downstream, so that the upstream loading is balanced [6–8]. Three-phase FACTs are another group of versatile devices that emerged in response to the increasing concern regarding power quality. They can provide reactive power support, voltage regulation, or harmonic compensation. With proper control, devices such as SVCs [5,9] and distribution static compensator (DSTATCOMs) can also achieve load unbalance compensation [10–13]. Although they can be used for power routing control, they require additional hardware. With the increasing integration of photovoltaics (PVs) and other distributed energy resources (DERs), three-phase power electronic interfaces have been proposed for compensation [14–16]. The back-to-back converters connecting asynchronous microgrids to the main distribution grid are also considered for unbalance compensation in [17]. Even though three-phase power electronic interfaces are attractive for IPPF control applications, their interphase power routing capability is limited as they are designed for balanced power operations. Unbalanced operations may induce unacceptable DC-link voltage fluctuation [18].

The contributions brought in the paper are three fold. Firstly, the IPPF theory is proposed for modeling the power flow behavior through single-phase devices connected between two different phases. The proposed model is applicable for constant impedance, constant current, and constant power elements connected in line-to-line or line-to-neutral configurations. The second contribution involves the development of control algorithms governing the IPPF of line-to-line single-phase SVCs and line-to-line single-phase power electronic interfaces. Additionally, the control is also applicable for the line-to-neutral SVCs and power electronic interfaces connected at the wye side of delta-wye transformers. Two control modes are proposed for line-to-line single-power electronic interfaces. The first mode provides the active and reactive power control of the devices. The second control mode enables the auxiliary controls including precise power injection and power routing control of two connected terminals. Lastly, two applications utilizing the coordinated IPPF controls for improving system operation and flexibility are provided.

The organization of this paper are outlined as follows: In Section 2, the IPPF model is proposed for modeling the power flow phenomena of single-phase elements. The model serves as the development framework for the IPPR control of line-to-line single-phase elements in Section 3. In Section 4, simulations are provided for demonstration and verification. Finally, Section 5 concludes the paper.

2. Interphase Power Flow via Single-Phase Element

In this section, general power flow phenomena of a single-phase element are investigated and modeled. For generality, single-phase elements are modeled as loads, which can be categorized into constant current, constant power, or constant impedance loads.

2.1. Interphase Power Flow

IPPF refers to power flow as the result of connecting a single-phase element between two terminals, a and b. The power flow of interest includes active power (P), reactive power (Q), and complex power (S). Active power is the power (measured in W) that is utilized by the element for real work. Reactive power is the power (measured in var) used to maintain electric and magnetic fields of the element. Complex power (measured in VA) is the sum of the active power and reactive power ($S = P + jQ$). The power flow of the single-phase element modeled in IPPF can be categorized as follows:

1. Complex, active, and reactive power absorption through each terminal: $\overline{S}_a$, P_a, Q_a, $\overline{S}_b$, P_b, Q_b,
2. Total power absorbed by the single-phase element: $\overline{S}_{ab}$, P_{ab} and Q_{ab},
3. The difference in the power absorption between two terminals: $\overline{S}_{\Delta ab}$, $P_{\Delta ab}$, and $Q_{\Delta ab}$. Positive indicates that the absorbed power on terminal a is higher than terminal b, i.e., $P_a > P_b$, or $Q_a > Q_b$.

IPPF from terminal a to b can be represented compactly as (1),

$$IPPF_{ab} = \begin{bmatrix} P_a & Q_a & P_b & Q_b & P_{ab} & Q_{ab} & P_{\Delta ab} & Q_{\Delta ab} \end{bmatrix}^t. \tag{1}$$

The relationship among the power in IPPF can be expressed mathematically as (2)–(5) and illustrated in Figure 1.

$$\overline{S}_{ab} = \overline{S}_a + \overline{S}_b, \tag{2}$$

$$\overline{S}_{\Delta ab} = \overline{S}_a - \overline{S}_b, \tag{3}$$

$$\overline{S}_a = \frac{\overline{S}_{ab}}{2} + \frac{\overline{S}_{\Delta ab}}{2}, \tag{4}$$

$$\overline{S}_b = \frac{\overline{S}_{ab}}{2} - \frac{\overline{S}_{\Delta ab}}{2}. \tag{5}$$

The power drawn by the single-phase element through each terminal ($\overline{S}_a$ or $\overline{S}_b$) consists of two parts. The first part ($0.5\,\overline{S}_{ab}$) is absorbed by the element, which is a half of the total power ($\overline{S}_{ab}$). The other part ($0.5\,\overline{S}_{\Delta ab}$), defined as "interphase power routing" (IPPR), is not absorbed by the load but routed through the load from terminal a to b. It is equal to half of the power difference between terminals. In the following subsections, the IPPF of different single-phase load types is investigated. The considered load models are as shown in Figure 2.

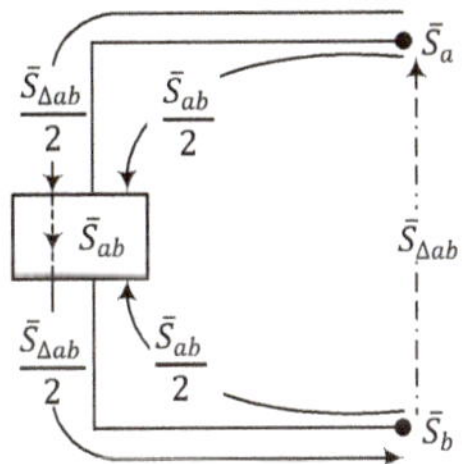

Figure 1. Interphase power flow model of a single-phase load connected across terminal a and b.

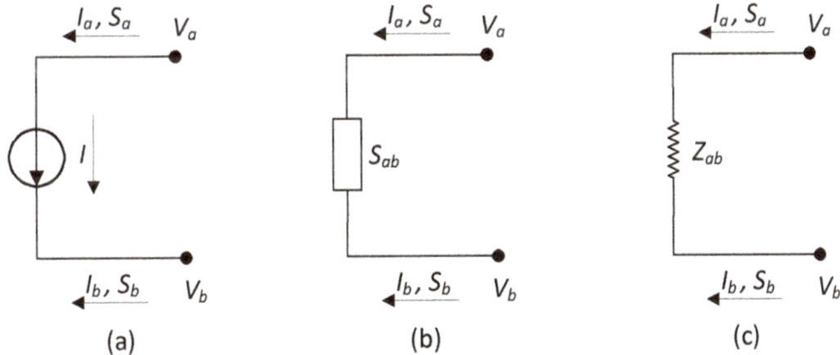

Figure 2. Physical models of single-phase constant (**a**) current, (**b**) power, (**c**) impedance loads.

2.2. Constant Current Load

Considering that the constant current load in Figure 2a draws in complex currents $\overline{I}_a$ and $\overline{I}_b$ from each terminal, respectively, the current flowing through the load is denoted as $\overline{I} = \overline{I}_a = -\overline{I}_b$. By the definition of complex power, the following relations hold:

$$\overline{S}_a = \overline{V}_a \overline{I}_a^* = \overline{V}_a \overline{I}^*, \tag{6}$$

$$\overline{S}_b = \overline{V}_b \overline{I}_b^* = -\overline{V}_b \overline{I}^*, \tag{7}$$

$$\overline{S}_{ab} = (\overline{V}_a - \overline{V}_b)\,\overline{I}^*, \tag{8}$$

$$\overline{S}_{\Delta ab} = (\overline{V}_a + \overline{V}_b)\,\overline{I}^*. \tag{9}$$

The steady-state complex voltages and currents in (6)–(9) can be expressed in rectangular coordinates with subscripts r and i representing the real and the imaginary parts, respectively, i.e., $\bar{I} = I_r + jI_i$ where I_r and I_i are the real and the imaginary parts of $\bar{I}$. As a result, (6)–(9) can be rewritten as (10),

$$
\begin{bmatrix} P_a \\ Q_a \\ P_b \\ Q_b \\ P_{ab} \\ Q_{ab} \\ P_{\Delta ab} \\ Q_{\Delta ab} \end{bmatrix} =
\begin{bmatrix}
V_{ar} & V_{ai} \\
V_{ai} & -V_{ar} \\
-V_{br} & -V_{bi} \\
-V_{bi} & V_{br} \\
V_{ar}-V_{br} & V_{ai}-V_{bi} \\
V_{ai}-V_{bi} & -V_{ar}+V_{br} \\
V_{ar}+V_{br} & V_{ai}+V_{bi} \\
V_{ai}+V_{bi} & -V_{ar}-V_{br}
\end{bmatrix}
\begin{bmatrix} I_r \\ I_i \end{bmatrix}.
\tag{10}
$$

Apart from depicting the relationship between the steady-state current and IPPF, (9) and (10) also show that constant current loads cause power differences; IPPR. The amount of IPPR is proportional to the current magnitude. Furthermore, constant current loads can trade off between real and reactive IPPR by varying the current angle as shown in (10). More active power and less reactive power is routed as current aligns more toward $\bar{V}_a + \bar{V}_b$.

2.3. Constant Power Load

The model of a constant power load consuming a total complex power of $\bar{S}_{ab}$ is shown in Figure 2b. By rearranging (10), the relationship between the constant power load and IPPF can be obtained in polar coordinate as (11), where V and θ denote voltage magnitude and angle respectively,

$$
\begin{bmatrix} P_a \\ Q_a \\ P_b \\ Q_b \\ P_{ab} \\ Q_{ab} \\ P_{\Delta ab} \\ Q_{\Delta ab} \end{bmatrix} =
\begin{bmatrix}
V_{ar} & V_{ai} \\
V_{ai} & -V_{ar} \\
-V_{br} & -V_{bi} \\
-V_{bi} & V_{br} \\
V_{ar}-V_{br} & V_{ai}-V_{bi} \\
V_{ai}-V_{bi} & -V_{ar}+V_{br} \\
V_{ar}+V_{br} & V_{ai}+V_{bi} \\
V_{ai}+V_{bi} & -V_{ar}-V_{br}
\end{bmatrix}
\begin{bmatrix} V_{ar}-V_{br} & V_{ai}-V_{bi} \\ V_{ai}-V_{bi} & -V_{ar}+V_{br} \end{bmatrix}^{-1}
\begin{bmatrix} P_{ab} \\ Q_{ab} \end{bmatrix},
$$

$$
= \frac{1}{V_{ab}^2}
\begin{bmatrix}
V_a\cos(\theta_a) & V_a\sin(\theta_a) \\
V_a\sin(\theta_a) & -V_a\cos(\theta_a) \\
-V_b\cos(\theta_b) & -V_b\sin(\theta_b) \\
-V_b\sin(\theta_b) & V_b\cos(\theta_b) \\
V_a\cos(\theta_a)-V_b\cos(\theta_b) & V_a\sin(\theta_a)-V_b\sin(\theta_b) \\
V_a\sin(\theta_a)-V_b\sin(\theta_b) & -V_a\cos(\theta_a)+V_b\cos(\theta_b) \\
V_a\cos(\theta_a)+V_b\cos(\theta_b) & V_a\sin(\theta_a)+V_b\sin(\theta_b) \\
V_a\sin(\theta_a)+V_b\sin(\theta_b) & -V_a\cos(\theta_a)-V_b\cos(\theta_b)
\end{bmatrix}
\begin{bmatrix} V_a\cos(\theta_a)-V_b\cos(\theta_b) & V_a\sin(\theta_a)-V_b\sin(\theta_b) \\ V_a\sin(\theta_a)-V_b\sin(\theta_b) & -V_a\cos(\theta_a)+V_b\cos(\theta_b) \end{bmatrix}
\begin{bmatrix} P_{ab} \\ Q_{ab} \end{bmatrix},
\tag{11}
$$

$$
= \frac{1}{V_{ab}^2} M \begin{bmatrix} P_{ab} \\ Q_{ab} \end{bmatrix} =
\underbrace{\begin{bmatrix} \frac{1}{2}P_{ab} \\ \frac{1}{2}Q_{ab} \\ \frac{1}{2}P_{ab} \\ \frac{1}{2}Q_{ab} \\ P_{ab} \\ Q_{ab} \\ 0 \\ 0 \end{bmatrix}}_{N}
+
\underbrace{\begin{bmatrix} uP_{ab} \\ uQ_{ab} \\ -uP_{ab} \\ -uQ_{ab} \\ 0 \\ 0 \\ 2uP_{ab} \\ 2uQ_{ab} \end{bmatrix}}_{U}
+
\underbrace{\begin{bmatrix} rQ_{ab} \\ -rP_{ab} \\ -rQ_{ab} \\ rP_{ab} \\ 0 \\ 0 \\ 2rQ_{ab} \\ -2rP_{ab} \end{bmatrix}}_{R},
$$

$$\text{where } M = \begin{bmatrix} V_a^2 - V_a V_b \cos(\theta_a - \theta_b) & V_a V_b \sin(\theta_a - \theta_b) \\ -V_a V_b \sin(\theta_a - \theta_b) & V_a^2 - V_a V_b \cos(\theta_a - \theta_b) \\ V_b^2 - V_b V_a \cos(\theta_b - \theta_a) & V_b V_a \sin(\theta_b - \theta_a) \\ -V_b V_a \sin(\theta_b - \theta_a) & V_b^2 - V_b V_a \cos(\theta_b - \theta_a) \\ V_{ab}^2 & 0 \\ 0 & V_{ab}^2 \\ V_a^2 - V_b^2 & 2V_a V_b \sin(\theta_a - \theta_b) \\ -2V_a V_b \sin(\theta_a - \theta_b) & V_a^2 - V_b^2 \end{bmatrix}, \; u = (V_a^2 - V_b^2)/2V_{ab}^2,$$

$r = V_a V_b \sin(\theta_a - \theta_b)/V_{ab}^2$ and $V_{ab} = ||\overline{V_a} - \overline{V_b}||$.

As shown in (11), the constant power load influence over IPPF can be decomposed into 3 components expressed as separate matrices. They can be interpreted as follows:

1. The first component, N represents the fundamental power distribution of a single-phase load. It shows that power is drawn equally from both terminals. This component can be considered the base power flow that does not involve IPPR.
2. The second component, U involves a factor u, which describes how a voltage magnitude unbalance causes an IPPR and an uneven power distribution in the base power flow described above. The power is drawn more from the terminal with the higher voltage magnitude.
3. The last component, R describes an IPPR as a result of the presence of the opposite power quantity in a single-phase load. In particular, the real power consumption of a load (P_{ab}) routes interphase reactive power of rP_{ab} from the lagging phase to the leading phase terminal. On the other hand, the reactive power consumption of a load (Q_{ab}) routes an interphase real power of rQ_{ab} from the leading phase to the lagging phase terminal.

The IPPF of a constant power load can be illustrated with the three components as in Figure 3.

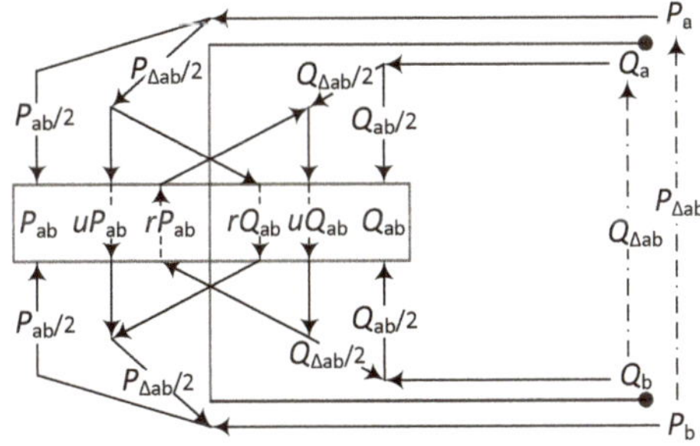

Figure 3. Interphase power flow of a constant power load.

Under the assumption that the voltage is balanced and the voltages of terminal a leads b by $120°$, u and r become 0 and $\frac{1}{2\sqrt{3}}$ respectively. Hence (11) simplifies to (12), which can be used to approximate the IPPF of a line-to-line connected constant power load.

$$\begin{bmatrix} P_a \\ Q_a \\ P_b \\ Q_b \\ P_{ab} \\ Q_{ab} \\ P_{\Delta ab} \\ Q_{\Delta ab} \end{bmatrix} \simeq \begin{bmatrix} 1/2 & 1/2\sqrt{3} \\ -1/2\sqrt{3} & 1/2 \\ 1/2 & -1/2\sqrt{3} \\ 1/2\sqrt{3} & 1/2 \\ 1 & 0 \\ 0 & 1 \\ 0 & 1/\sqrt{3} \\ -1/\sqrt{3} & 0 \end{bmatrix} \begin{bmatrix} P_{ab} \\ Q_{ab} \end{bmatrix}. \tag{12}$$

2.4. Constant Impedance Load

A constant impedance load denoted Z_{ab} can be modeled as in Figure 2c with an admittance of $G_{ab} + jB_{ab}$. IPPF in (11) can be rewritten as (13) where M is as defined in (11).

$$IPPF_{ab} = M \begin{bmatrix} G_{ab} \\ -B_{ab} \end{bmatrix}. \tag{13}$$

Similar to the constant power load analysis, (13) describes the precise IPPF regardless of the terminal voltages. However, with a balanced voltage assumption (the voltage angle of phase a leads phase b by $120°$), (13) is simplified to (14).

$$\begin{bmatrix} P_a \\ Q_a \\ P_b \\ Q_b \\ P_{ab} \\ Q_{ab} \\ P_{\Delta ab} \\ Q_{\Delta ab} \end{bmatrix} \simeq V_{ab}^2 \begin{bmatrix} 1/2 & -1/2\sqrt{3} \\ -1/2\sqrt{3} & -1/2 \\ 1/2 & 1/2\sqrt{3} \\ 1/2\sqrt{3} & -1/2 \\ 1 & 0 \\ 0 & -1 \\ 0 & -1/\sqrt{3} \\ -1/\sqrt{3} & 0 \end{bmatrix} \begin{bmatrix} G_{ab} \\ B_{ab} \end{bmatrix}. \tag{14}$$

The influence of the constant impedance load is similar to that of the constant power load. The conductance of the load draws real power equally from both terminals while routing interphase reactive power ($\frac{V_{ab}^2 G_{ab}}{2\sqrt{3}}$) from lagging to leading terminal. On the other hand, the inductive load (negative susceptance) draws reactive power equally from both phases and routes interphase real power ($\frac{-V_{ab}^2 B_{ab}}{2\sqrt{3}}$) from leading to lagging terminal.

2.5. Line-to-Neutral Load

In this section, a line-to-neutral load is analyzed as a special case of a line-to-line load when one of its terminals is connected to neutral. Without loss of generality, the load is modeled as a constant power load and terminal b is connected to neutral (denoted n) at which the voltage is zero. As a result, expression (11) simplifies to (15).

$$\begin{bmatrix} P_a \\ Q_a \\ P_n \\ Q_n \\ P_{an} \\ Q_{an} \\ P_{\Delta an} \\ Q_{\Delta an} \end{bmatrix} = \underbrace{\begin{bmatrix} \frac{1}{2}P_{ab} \\ \frac{1}{2}Q_{ab} \\ \frac{1}{2}P_{ab} \\ \frac{1}{2}Q_{ab} \\ P_{ab} \\ Q_{ab} \\ 0 \\ 0 \end{bmatrix}}_{N} + \underbrace{\begin{bmatrix} \frac{1}{2}P_{ab} \\ \frac{1}{2}Q_{ab} \\ -\frac{1}{2}P_{ab} \\ -\frac{1}{2}Q_{ab} \\ 0 \\ 0 \\ P_{ab} \\ Q_{ab} \end{bmatrix}}_{U} + \underbrace{\begin{bmatrix} 0 \\ 0 \\ 0 \\ 0 \\ 0 \\ 0 \\ 0 \\ 0 \end{bmatrix}}_{R}. \tag{15}$$

When the neutral voltage is zero, factors u and r in (11) become 0.5 and 0 respectively. This indicates that power routing control becomes ineffective as R component does not affect IPPF. Moreover, at the neutral terminal, the IPPR caused by U ($-0.5\,P_{ab}$ and $-0.5\,Q_{ab}$) is drawn back to the load by the N component ($0.5\,P_{ab}$ and $0.5\,Q_{ab}$). This results in no power flow at the neutral terminal. However, some power can leak to the neutral network and small interphase power routing is possible when the neutral voltage is non-zero. Nevertheless, since neutral voltage in practice can be neglected, the line-to-neutral single-phase loads do not produce an IPPR between its phase terminal and neutral.

2.6. Delta-Wye Transformer

The circuit diagram of a delta-wye transformer (Dy1) according to ANSI/IEEE C57.12 is shown in Figure 4. The secondary windings of phase a-n, b-n, and c-n are connected directly across the primary windings of phase A-C, B-A, and C-B, respectively. As a result, line-to-neutral loads on the secondary side are perceived effectively as line-to-line loads on the primary perspective. The IPPF at the delta side of a delta-wye transformer caused by a phase a load at the wye side can be calculated with (12).

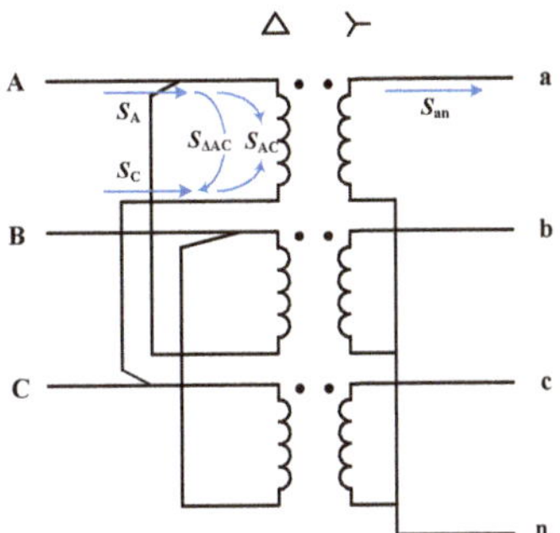

Figure 4. A circuit diagram of a delta-wye transformer (Dy1) and the IPPF at the delta winding as a result of a phase a load on the wye winding.

3. Interphase Power Flow Control

In the previous section, the IPPF models for different load types were developed, showing that the line-to-neutral elements behind delta-wye transformers and line-to-line single-phase elements can be utilized for routing power between terminals. The reactive power of the elements can be utilized to route interphase real power while the real power can be used for routing interphase reactive power. In this section, control methodologies are proposed for controlling line-to-line SVCs and power electronic interfaces to achieve desirable IPPR and IPPF. The controls are developed for the line-to-line elements for simplicity; however, it is also applicable for line-to-neutral single-phase elements connected at the wye side of a delta-wye transformer.

3.1. Line-to-Line Static Var Compensator (SVC)

SVCs are shunt devices consisting of reactance bank which may employ either capacitors or inductors. In general, SVCs are deployed to provide reactive power support at selected locations. To provide the desired reactive power support, the corresponding reactive power or reactance is determined and SVCs then adjust their capacitance or inductance appropriately.

From the IPPF control perspective, SVCs are considered a constant impedance load with uncontrollable and negligible resistance. According to (13) and (14), interphase reactive power routing achievable ($Q_{\Delta ab}/2$) would also be uncontrollable and small. Hence, the IPPR control of SVCs will focus only on real power routing. To achieve a desired interphase real power routing, $P_{\Delta ab}/2$, the appropriate reactive power and reactance of SVCs can be computed precisely by using (16) and (17):

$$Q_{ab} = \frac{V_{ab}^2 P_{\Delta ab} - (V_a^2 - V_b^2) P_{ab}}{2 V_a V_b \sin(\theta_a - \theta_b)}, \tag{16}$$

$$B_{ab} = -\frac{P_{\Delta ab} - (V_a^2 - V_b^2) G_{ab}}{2 V_a V_b \sin(\theta_a - \theta_b)}. \tag{17}$$

The reactive power and reactance can also be estimated with balanced a voltage assumption. From (12) and (14), (18) and (19) can be obtained as follows:

$$Q_{ab} = \sqrt{3}P_{\Delta ab},\tag{18}$$

$$B_{ab} = -\frac{\sqrt{3}P_{\Delta ab}}{V_{ab}^2}.\tag{19}$$

3.2. Line-to-Line Power Electronic Interface

This subsection proposes an online IPPR control for a power electronic interface. The model of a power electronic interface connected to the grid in line-to-line configuration is shown in Figure 5. The power electronic interface consists of an inverter, a filter and a DC link, which is connected to DC power source such as PV, energy storage or an electric vehicle. Unlike SVCs, power electronic interfaces utilize power electronic gates and PWM techniques.

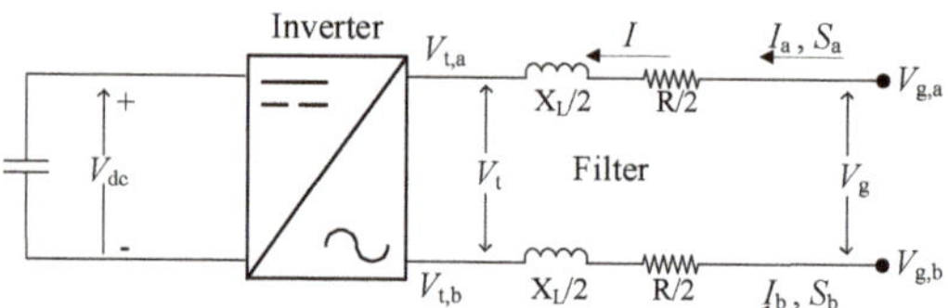

Figure 5. Physical model of a single-phase line-to-line power electronic interface.

The overview of the control algorithm is shown in Figure 6. The proposed controller consists of phase lock loop (PLL) and three hierarchical control loops. The PLL unit serves as the observer for the system that uses the measurement of physical states to estimate other states of interest. The estimated states will be prompted for the controllers to utilize. The outermost controller loop is the IPPF control loop, whose main task is to calculate the corresponding target AC current flow from a desired IPPF. The current control loop's responsibility is to determine the appropriate voltage at the inverter terminal. Finally, the voltage control loop issues the control signal for the power electronic gates in the inverter to realize the target voltage set by the current controller.

In the practical implementation, another control for the DC link voltage should also be considered. However, since this is not relevant for the main functionality, the DC link voltage control is omitted. In this work, the DC-link voltage is assumed to be perfectly maintained by a constant voltage source.

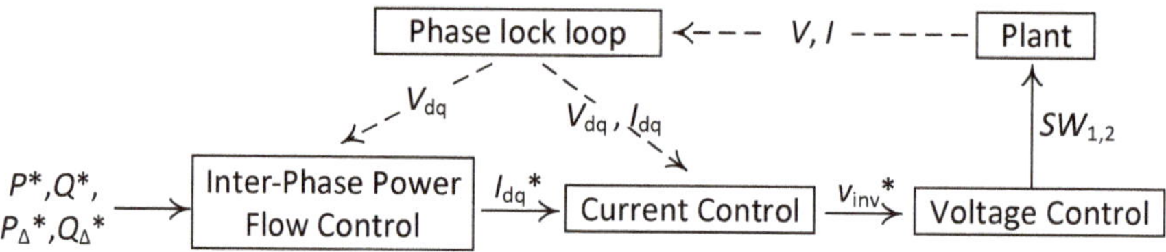

Figure 6. Interphase power flow control blocks of power electronic interfaces.

3.2.1. Phase Lock Loop (PLL)

Since expression (10) relating current and IPPF can be regarded as a representation of a synchronous rotating reference frame or *d-q* domain, the referencing frame is chosen as the coordinate system for developing the control hence forth. As a result, it is necessary for the controller to have a means for extracting the states of interest in such coordinate system. This can be achieved by utilizing second-order-general integrator quadrature signal generator (SOGI-QSG) to generate quad delay signals [19] and using the Park Transform to convert the single phase stationary coordinates to the *d-q* reference coordinates. However, since the outputs of SOGI-QSG are instantaneous, the values obtained after Park transformation are amplitudes. The RMS values can be calculated by dividing the outputs by $\sqrt{2}$.

In this section, d and q components are considered as real and imaginary parts respectively. Furthermore, PLL will regard line-to-line voltage at grid side (V_g) as the reference for the d-q rotating frame. In particular, the rotating d-q frame shall synchronously align with V_g such that the q component of V_g will be adjusted to zero. A simple PI controller achieves alignment. By using the aligned d-q rotating frame as a reference, other properties of interest in the d-q frame can be determined via Park Transform. The control block of SOGI-QSG is as shown in Figure 7. The closed loop transfer functions of V_d and V_q are as shown in Equations (20) and (21). It can be observed that the gain at ω of $\frac{v_d}{v}$ is 1 while the gain of $\frac{v_q}{v}$ is $-j$ (90° delay of the same magnitude). The gain k determines the closed-loop bandwidth.

$$\frac{v_d}{v} = \frac{k\omega s}{s^2 + k\omega s + \omega^2}.$$

(20)

$$\frac{v_q}{v} = \frac{k\omega^2}{s^2 + k\omega s + \omega^2}.$$

(21)

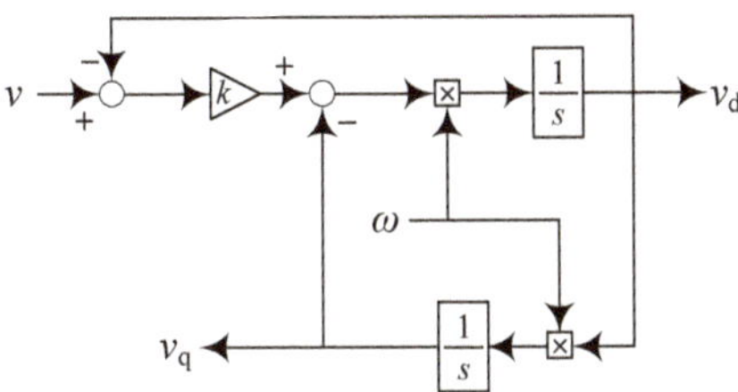

Figure 7. Second-order-general integrator quadrature signal generator control block.

3.2.2. Interphase Power Flow Control Loop

The main purpose of IPPF controller is to determine the target current flowing in the AC side to attain a desired IPPF characteristic with respect to the real time system voltage. This can be accomplished by using the inverse of (11), but this method introduces several challenges. Firstly, as the rank of the transfer matrix is two, only two power quantities can be chosen for control while the other three dependent power quantities would be contingently determined and indirectly controlled. The importance of each power quantity in the control perspective is as follows:

1. P_{ab} determines the active power transfer between the DC side and AC side. An inappropriate setting of P_{ab} would result in the increase and decrease of the DC link voltage.
2. Q_{ab} determines the total reactive power absorption which may be dispatched from a centralized controller.
3. $P_{\Delta ab}$ and $Q_{\Delta ab}$ relate to real and reactive power routing control.
4. P_a, P_b, Q_a and Q_b may be selected for precise real and reactive power absorption at the terminal.

The second challenge is that although there are four candidate control parameters, (12) suggests that they should not be chosen arbitrarily. Particularly, under balanced voltage conditions, interphase real power routing ($P_{\Delta ab}$) and total reactive power (Q_{ab}) are dependent and interphase reactive power routing ($Q_{\Delta ab}$) and total real power (P_{ab}) are dependent. Thus, the two power quantities in each pair should not be selected simultaneously.

Considering the variable importance and dependency, two operation modes and the corresponding controls are proposed:

1. In the first operating mode in which the set point of P_{ab} and Q_{ab} are given, P_{ab} and Q_{ab} should be selected as the control parameters. When Q_{ab} is small, the real power should be roughly equally distributed between the two phases.

2. In the second mode, power electronic interfaces are utilized for ancillary services such as IPPF control or precise power absorption or injection. In this mode, two power quantities of IPPF can be selected for control. To control active and reactive IPPR, $P_{\Delta ab}$ or $Q_{\Delta ab}$ can be selected. For controlling the precise power injection at a connected terminal, P_a, P_b, Q_a and Q_b can be selected. However, neither $P_{\Delta ab}$ and Q_{ab}, nor $Q_{\Delta ab}$ and P_{ab} should be chosen simultaneously. After the control parameters have been determined, the corresponding P_{ab} and Q_{ab} to achieve the desired ancillary service can be calculated by using Equation (11). Those values should be limited within the inverter power rating. Moreover, the AC-DC active power balance should be enforced to ensure stable DC link voltage.

Regardless of the operation modes, after P_{ab}, Q_{ab} have been determined, Equation (10) can be used to calculate the corresponding target currents, I_d^* and I_q^*.

3.2.3. Current Control Loop

The purpose of the current controller is to determine the target instantaneous terminal voltages at the inverter AC side to realize the target current issued by the IPPR controller given the current system voltage. Since the component in the d-q reference is readily available, the control is developed using the double synchronous reference frame (DSRF) control scheme. From Figure 5, the following system dynamics can be derived.

$$\begin{cases} V_{g,a} - V_{t,a} = \dfrac{R}{2}I_a + \dfrac{L}{2}\dfrac{dI_a}{dt}, \\[2mm] V_{g,b} - V_{t,b} = \dfrac{R}{2}I_b + \dfrac{L}{2}\dfrac{dI_b}{dt}. \end{cases} \tag{22}$$

These translate to

$$V_g - V_t = RI + L\frac{dI}{dt}. \tag{23}$$

Also in the d-q domain as

$$\begin{cases} L\dfrac{dI_d}{dt} = V_{g,d} - V_{t,d} - RI_d + \omega L I_q, \\[2mm] L\dfrac{dI_q}{dt} = V_{g,q} - V_{t,q} - RI_q - \omega L I_d. \end{cases} \tag{24}$$

Following DSRF control scheme, the controller employs 2 PI controllers, one each for d and q reference frame. From the system dynamics, it can be observed that the coupling term can be decoupled by using backward compensation. Furthermore, forward compensation is utilized to cancel out the constant terms ($V_{g,d}$ and $V_{g,q}$) to improve system dynamics. The control block diagram is as shown in Figure 8. After the target AC inverter terminal voltage V_d and V_q are obtained, the inverse Park transform is applied to convert the voltage in d-q coordinates to the instantaneous coordinates.

3.2.4. Voltage Control Loop

After receiving the target instantaneous voltage magnitude, the voltage controller's responsibility is to generate the corresponding power electronic gate control signals. The voltage controller is similar to other general sinusoidal voltage controllers, for example, employing unipolar sinusoidal pulse width modulation.

4. Simulation and Verification

In this section, the IPPF model developed in Section 2 and the IPPF control proposed in Section 3 are verified and demonstrated by the simulation of two systems.

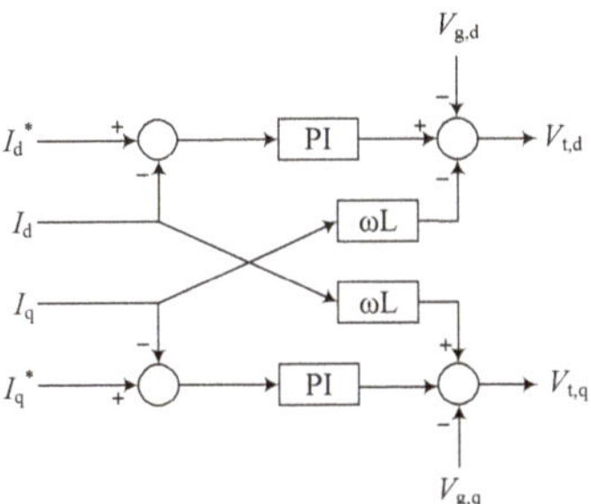

Figure 8. Double synchronous reference frame current control block.

4.1. Multimicrogrid System Simulation

The first simulation is performed in PSCAD to demonstrate the application of IPPF control for power routing in a multimicrogrid system.

4.1.1. System Description

The system consists of three grid-tied microgrids. The overview of the circuit and the PSCAD implementation are as shown in Figures 9 and 10, respectively. The first microgrid (MC1) is connected to the primary feeder through a 4.16 kV/208 V wye-wye transformer. This microgrid has 2 energy storage devices (ES1 and ES2) and a PV (PV1) connected across phases A-N, B-C, and A-B, respectively. The second microgrid (MC2) is served by a 4.16 kV/208 V delta-wye transformer. There are a PV device (PV3), and two ESs (ES3 and ES4) on phase b, c, and a, respectively. The third microgrid (MC3) is tapped from phase C on primary through a single-phase 2.4 kV/120–240 V transformer. A PV (PV2) provides distributed generation on this microgrid. For simplicity, loads are neglected, and all single-phase device ratings are 50 kVA. The initial power flow of the circuits is as shown in Figure 10. All power is provided from the primary feeder for the line and transformer losses.

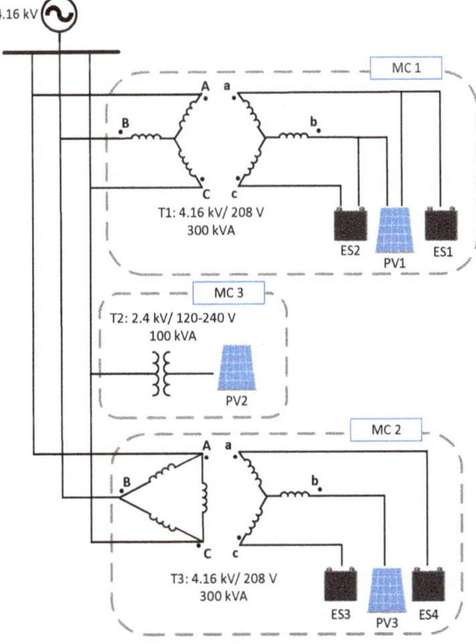

Figure 9. Multi-microgrid system.

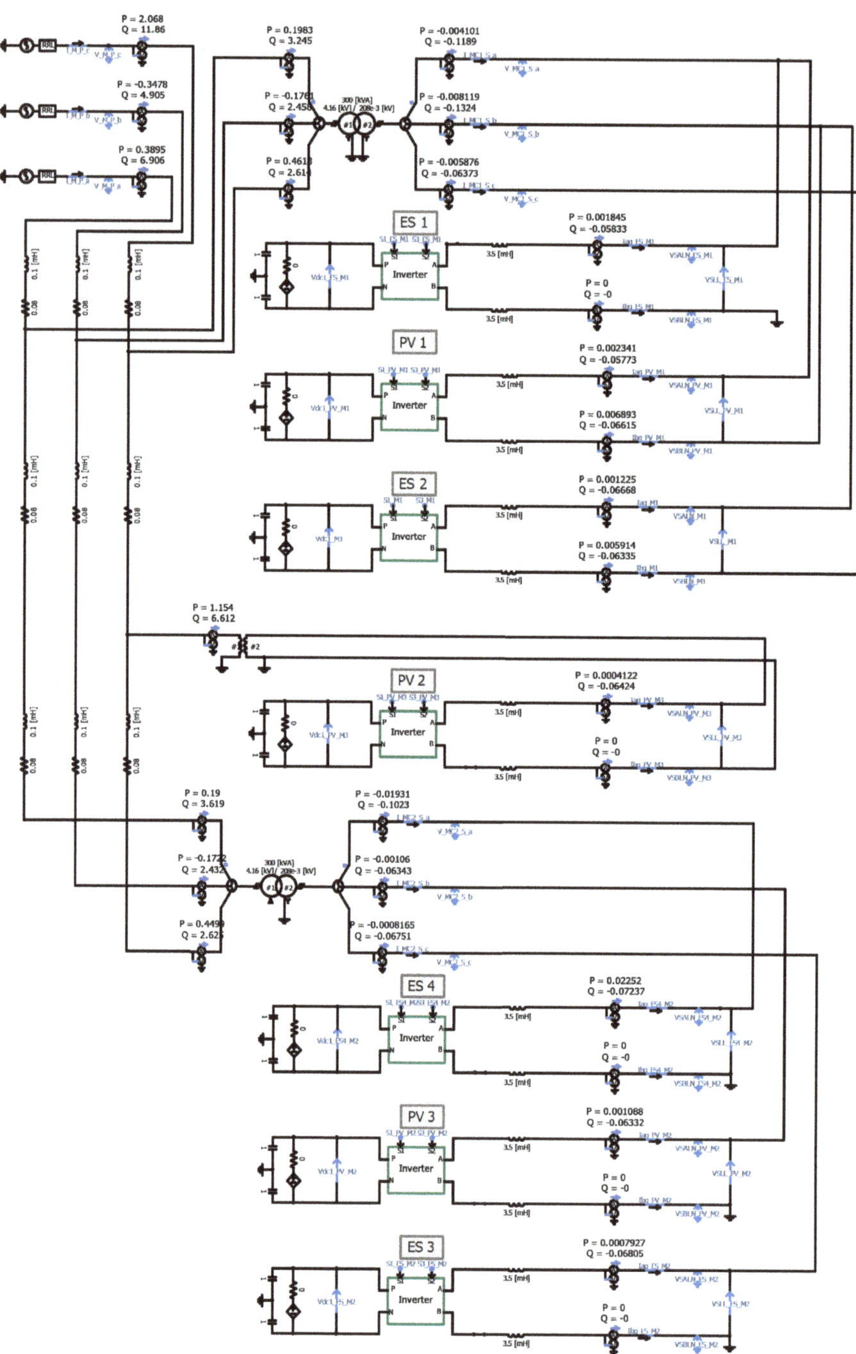

Figure 10. PSCAD model of the multimicrogrid systems. The multimeters display the initial active and reactive power flow (in kW and kvar) of the circuit with no power injection from PVs and ESs.

4.1.2. Interphase Power Flow Control

Three control actions are performed in a chronological order to demonstrate IPPF application:

1. At 0.4 s, 10 kW from PV3 (on phase B) is routed to charge ES3 (phase C). Additionally, the steady state active and reactive powers are provided entirely by the devices in the microgrid, making MC2 self-sufficient. The active and reactive power flow required for achieving this power routing can be calculated using the approximate IPPF model (12). The theoretical power flow at the delta-wye transformer is as shown in Figure 11.
2. At 0.6 s, 10 kW from PV2 in MC3 (phase C) is routed to charge ES2 in MC1 (on phase B-C). ES2 injects 17.3 kvar so that the active power is drawn solely from phase C.
3. At 0.8 s, 10 kW from PV3 in MC2 (phase B) and 10 kW from PV1 in MC1 (phase A-B) are routed to charge ES1 in MC1 (phase A). The total reactive power required for IPPF control can be calculated using (12). The reactive power burden for IPPF control can be distributed evenly among the PVs, requiring 17.3 kvar from each.

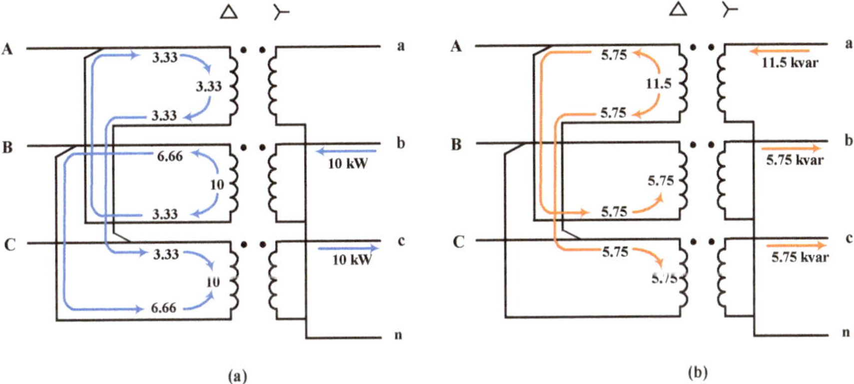

Figure 11. Theoretical interphase active (**a**) and reactive (**b**) power flow at the delta-wye transformer (Dy1) of MC2 as a result of routing 10 kW from phase B to C at the secondary circuit. The active and reactive power required for IPPF control are provided solely from the secondary circuit.

4.1.3. Result

Real powers at the main feeder head, the secondary side of the delta-wye transformer in MC2, the primary A-B winding of delta side of the delta-wye transformer in MC2, and the secondary side of the Wye-Wye transformer in MC1, and PV1 are shown in Figures 12–16, respectively. Figure 17 shows the steady-state power flow of multimicrogrid circuit after the all power routing. Additionally, Figure 18 is the steady-state power flow of the delta-wye transformer (T3) in MC2 after the first power routing action. The results show that the proposed IPPF control is effective in routing power between microgrids for achieving the desired objectives.

After the first control action at 0.4 s, active power is successfully routed from PV3 (phase B) to ES3 (phase C). Figure 18 shows the simulated power flow of MC 2, which matches with the theoretical power flow in Figure 11. It can be observed that the real and reactive power utilized for IPPF control is provided from within the microgrid as the power flow at the primary side of T3 is similar to the initial power flow in Figure 10. Furthermore, the active power in the secondary circuit is routed between phases successfully through the primary delta winding as desired according to Figure 11. Figure 14 shows the simulated active power injection from the A-B winding of the primary side of T3. Before the first IPPF control action is applied (before 0.4 s), it can be observed that small active power (0.577 kW) is routed from phase A to phase B due to the reactive power loss in the transformer winding. After the first IPPF control action, 9.998 kW from phase b in the secondary circuit is transferred to the primary A-B winding of T3. The power is then split into two portions, 3.308 and 6.69 kW. Furthermore, 6.69 kW is injected from the A-B winding to the B-C winding through phase B terminals. The other portion, 3.308 kW, is injected from the A-B winding through phase A terminal and then routed through A-C winding into B-C winding. Finally, the total active power in B-C winding of the delta side is transferred to the c phase in the secondary circuit.

After 0.6 s, the power from PV2 (phase C) in MC3 is routed to charge ES2 (phase B-C) in MC1. The reactive power of ES2 is controlled for power routing so that the active power is drawn through phase C only as observed in Figures 15 and 17. IPPF control enables single-phase elements to regulate their power drawn or injected at each terminal.

After 0.8 s, active power from PV1 and PV3 is routed to charge ES1. Since IPPF control for active power routing utilizes reactive power, the main active power operation of the participant is not interrupted. PV1 and PV3 can generate the active power while also utilizing their reactive power for IPPF application. Furthermore, the reactive power burden can also be shared among participant devices in MC1 and MC2.

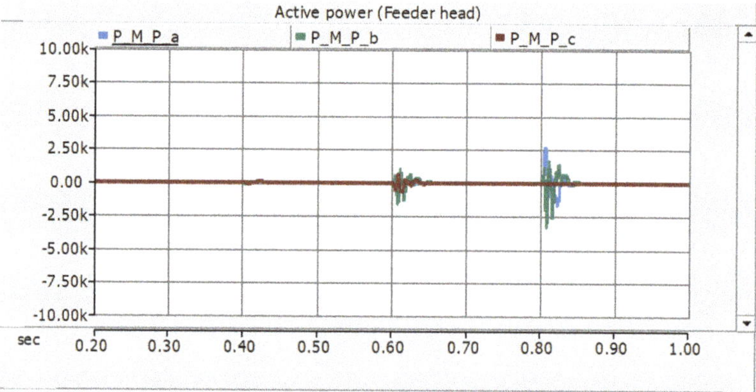

Figure 12. Feeder head active power shows that the power is shared between the multimicrogrid system effectively without requiring the power from the feeder head.

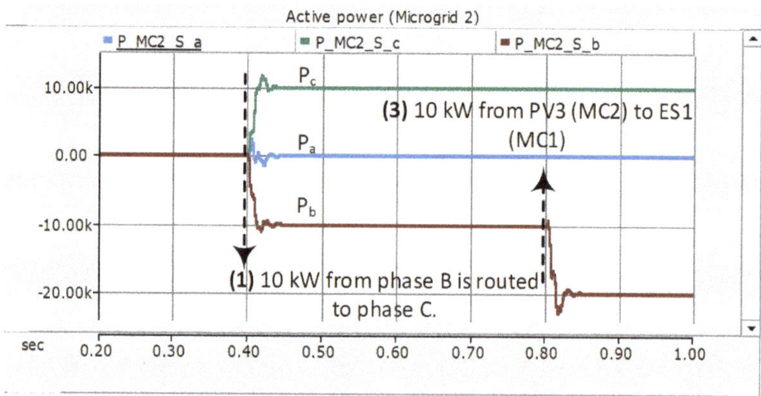

Figure 13. Active power at the wye side at the delta-wye transformer (Dy1) of MC2. Active power in phase A, B and C are labeled as P_a, P_b, P_c, respectively.

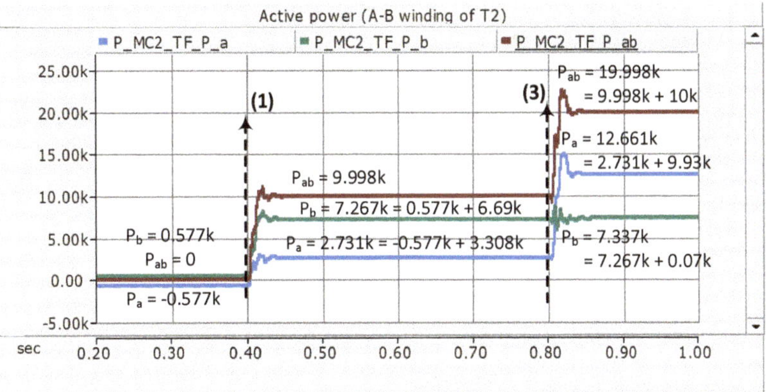

Figure 14. Active power across A-B winding at the delta side of T3. Active power injection to terminal A, B, and total power are labeled as P_a, P_b, P_{ab}, respectively.

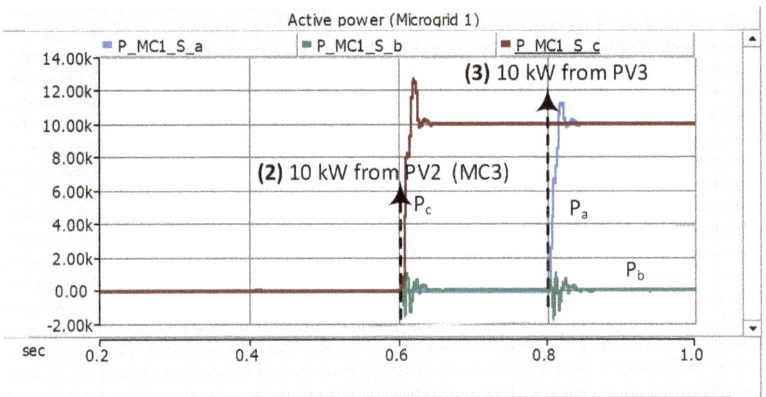

Figure 15. Active power at the secondary side at the wye-wye transformer of MC1. Active power in phase A, B and C are labeled as P_a, P_b, P_c, respectively.

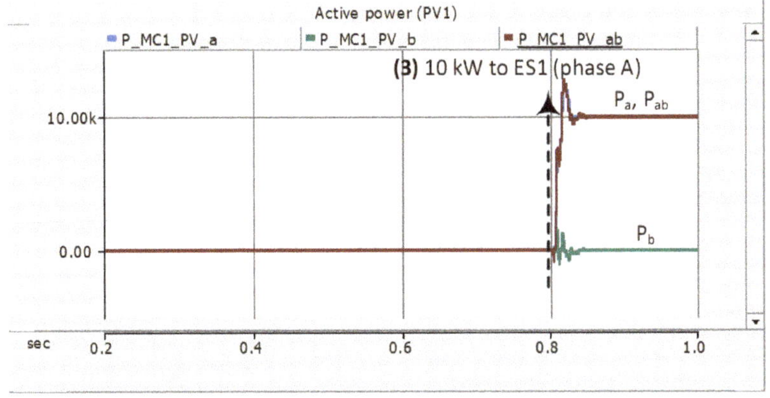

Figure 16. Active power of PV1 in MC1. Active power injection to phase A, B, and total power are labeled as P_a, P_b, P_{ab}, respectively.

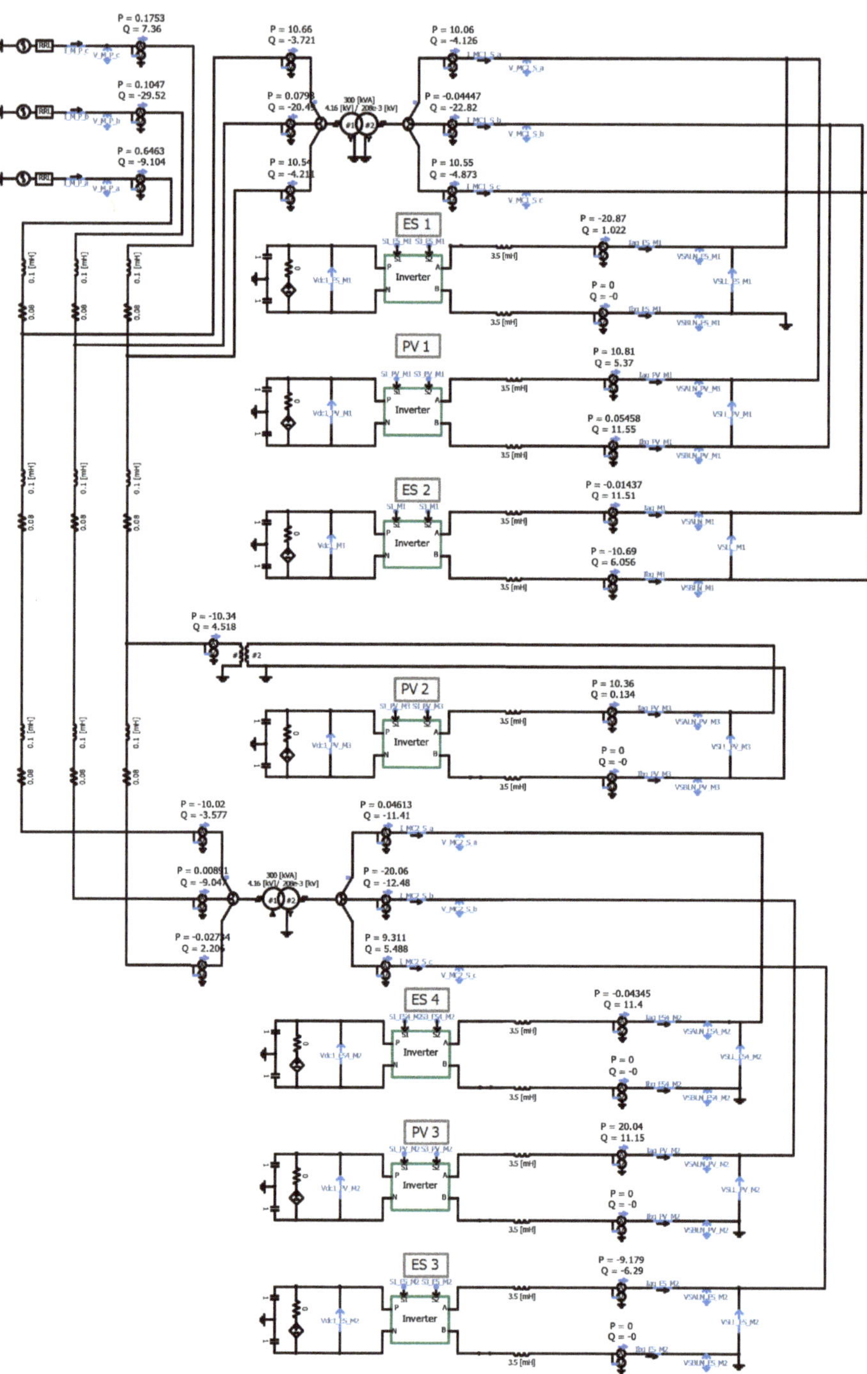

Figure 17. PSCAD simulation shows the active and reactive power flow (in kW and kvar) of the multimicrogrid system after all power routing actions. (1) Within MC2, 10 kW from PV3 (phase B) is routed to ES3 (phase C), (2) 10 kW from PV2 in MC3 (phase C) is routed to charge ES2 in MC1 (on phase B-C), and (3) 10 kW from PV3 in MC2 (phase B) and 10 kW from PV1 in MC1 (phase A-B) are routed to charge ES1 in MC1 (phase A).

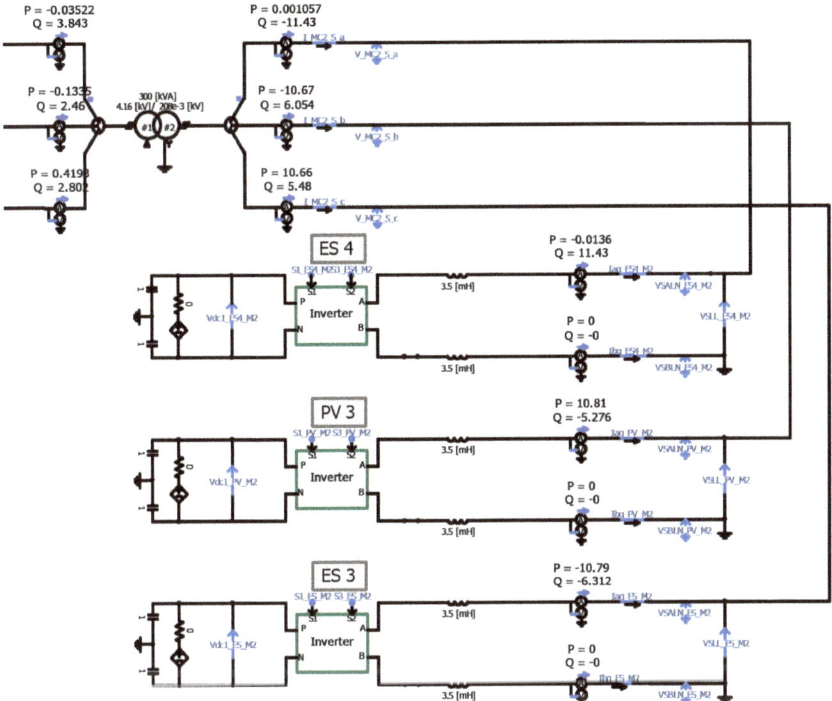

Figure 18. PSCAD simulation shows the active and reactive power flow (in kW and kvar) of Microgrid 2 after the first power routing action. 10 kW from phase B is routed to C at the secondary circuit. The active and reactive power required for IPPF control are provided solely from the secondary circuit.

The results in this simulation verify the IPPF models and controls developed in Section II and III. Furthermore, the simulation demonstrates the potential of IPPF control application for improving the system operation and equipment utilization of a multiphase microgrid and multimicrogrid systems. Within a microgrid, the generation of a phase can be routed to the designated equipment in another phase. IPPF control also enables the resources of multiple microgrids to be shared effectively despite the different phase connections from the primary. In addition, the simulation shows that real power routing as an ancillary function does not interrupt the main real power operation of the power electronic interfaces. Lastly, the reactive power burden required for IPPF control can be shared among the participating devices in different microgrids.

4.2. Distribution Feeder Simulation

In this simulation, a simple IPPF control is applied to a distribution circuit for balancing the active power at the feeder head. The control effectiveness and the impacts on system voltage unbalance and loss are evaluated. The simulation is performed in the Matlab—OpenDSS environment.

4.2.1. System Detail

The simulation circuit is a 12 kV distribution feeder of a utility. The one-line diagram is plotted in Figure 19 using GridPV tool [20]. This feeder is served by a 25 MVA, 69/12 kV, wye-wye connected substation transformer. The circuit has unbalanced loads of 2795, 3216, 3260 kW, in phases A, B, and C, respectively. Additionally, there are 215 single-phase

PVs generating a total power of 1806, 916, 716 kW, in phases A, B, and C, respectively. With loads and PVs connected, the total power at the feeder head is 953.4 + j 692, 2369.8 + j 1104.5, 2572.6 + j 1152.6 kVA as shown in Figure 20a. It can be observed that this power unbalance is the result of the lightly loaded conditions and high PV penetration in Phase A in comparison to the other phases. The distribution of the voltage unbalance of all buses is shown in Figure 21 under "base case" scenario. Voltage unbalance of most buses is in the range 0.8–1.6% with an average of 1.2%. A few buses have voltage unbalance below 0.8%. These buses are located near the substation where voltage is balanced. In general, it can be observed that the farther away from the feeder head, the higher the voltage unbalance.

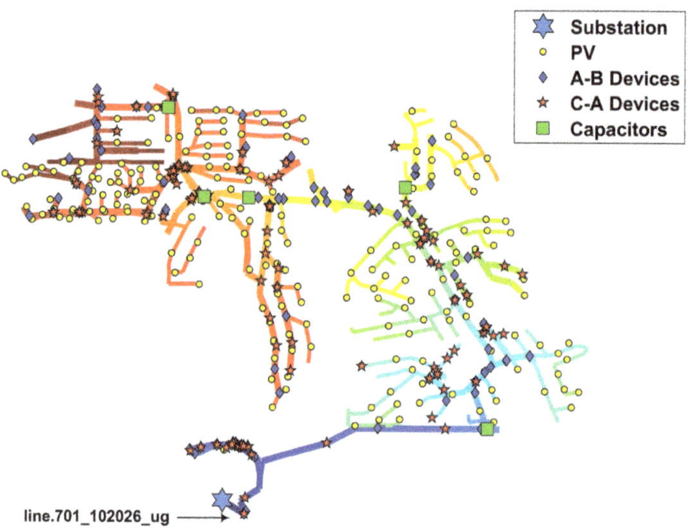

Figure 19. One-line diagram of the distribution circuit shows the locations of the PVs (PVs), line-to-line power electronic interfaces connected across A-B phase (A-B devices) and C-A phase (C-A devices), and capacitor banks.

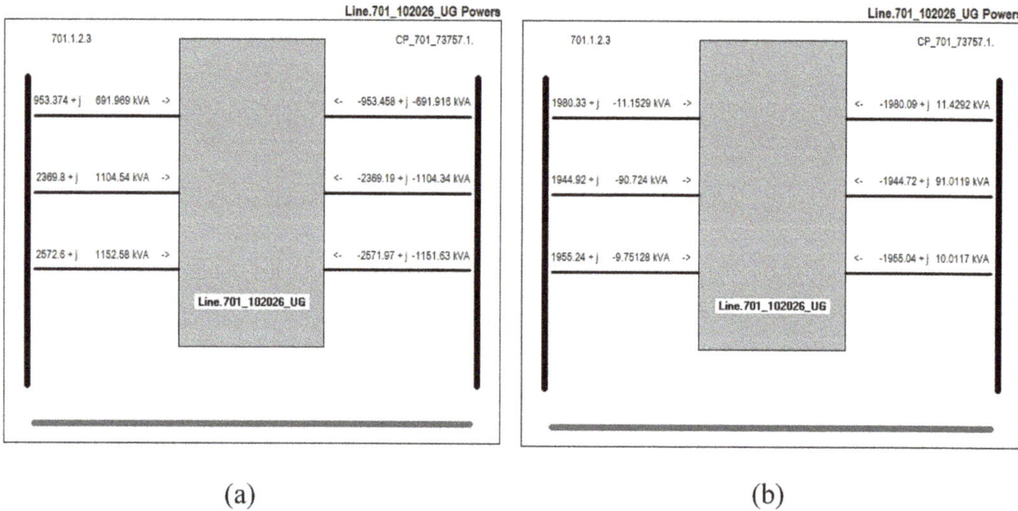

(a)   (b)

Figure 20. Substation active and reactive power flow in (**a**) base case, and (**b**) Case 2.

For IPPF control application demonstration purpose, it is assumed that there are 130 single-phase power electronic interfaces connected across each A-B and A-C phase. These devices can inject or absorb 12 kvar for IPPF control application. In addition, 5150-kvar capacitor banks across A-C phase are assumed to be connected at the primary feeder. These capacitor banks can be used in conjunction with the power electronic interfaces to share the IPPF control burden.

4.2.2. Interphase Power Flow Control

Three scenarios are considered for evaluating the IPPF control. The first scenario, "base case", is the original feeder without IPPF control. This scenario will serve as the baseline for comparison.

In the second scenario, denoted "Case 1", a simple IPPF control is implemented for routing active power from phase A to phase B and C so that the active power is balanced at the feeder head. The total reactive power requirement for IPPF application is calculated using (12). This reactive power burden is then shared evenly among all IPPF control devices; 10.4, −10.7 and 150 kvar is required from each A-B and A-C phase IPPF control device and capacitor bank, respectively.

The third scenario, denoted "Case 2", has IPPF control implemented the same as Case 1. In addition, the existing PVs are utilized for power factor correction so that power factor is unity at the feeder head. The reactive power requirement for power factor correction is calculated from the result of Case 1 and shared evenly among the PVs.

4.2.3. Simulation Results

The active and reactive power at the feeder head and the voltage unbalance distribution of all scenarios are summarized in Table 1 and Figure 21, respectively. In Case 1, the power at the feeder head becomes 1988.7 + j 366, 1906.1 + j 1767.3, and 1926.3 − j 32.1 kVA in phase A, B, and C, respectively. The system loss is reduced by 4%. The active power becomes balanced as expected after IPPF control. The reactive power is reduced in phase A and C, while increased in phase B. The reactive power remains unbalanced as it is not controlled and utilized for active power routing.

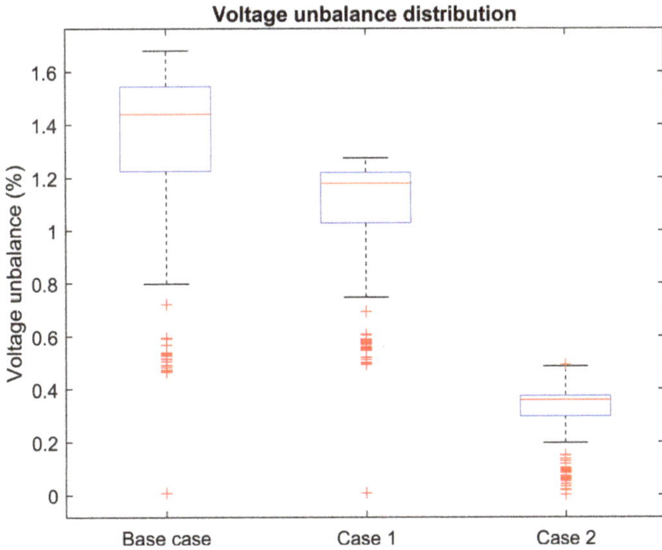

Figure 21. Voltage unbalance distribution of base case, Case 1 with IPPF control, and Case 2 with IPPF control and PF correction.

Table 1. Active and reactive power at the feeder head of base case, Case 1 and Case 2.

Scenarios	Feeder Head Power					
	Active Power (kW)			Reactive Power (kvar)		
	P_a	P_b	P_c	Q_a	Q_b	Q_c
Base case	953.4	2369.8	2572.6	692	1104.5	1152.6
Case 1	1988.7	1906.1	1926.3	366	1767.3	−32.1
Case 2	1980	1944.9	1955.2	−11.2	−90.7	−9.8

As observed in Figure 21, the average voltage unbalance is reduced from base case to 1%. Furthermore, significant voltage unbalance reduction is observed for the buses further away from the substation (voltage unbalance greater than 50th percentile). This demonstrates the effectiveness of performing unbalanced compensation in a distributive manner. Power unbalance is compensated throughout the system and voltage unbalance is improved throughout the system rather than a few locally.

In Case 2, the substation power after power routing is as shown in Figure 20b. Both active and reactive power at the feeder head becomes balanced: $1980 - j\,11.2$, $1944.9 - j\,90.7$, $1955.2 - j\,9.8$ kVA in phase A, B, and C, respectively. Significant improvement can be observed in system loss and voltage unbalance. System loss is reduced by 14% while the average voltage unbalance is reduced to 0.2%. Compared to Case 1, balance in both active and reactive power is achieved by using the line-to-neutral PVs to complement the IPPF control devices. The line-to-neutral PVs help compensate the reactive power in each phase which is not controlled in the IPPF control.

The simulation in this section demonstrates the effectiveness of the simple IPPF control in balancing the active power at the feeder head. The resulting system shows improvement in both voltage unbalance and system loss. The voltage unbalance of all buses show improvement rather than a few locally as the unbalance compensation is performed in a distributed manner throughout the circuit. Moreover, the line-to-neutral devices can be utilized to control or compensate the reactive power which is regulated in the active power routing. With both IPPF control and power factor correction, significant voltage unbalance and system loss is observed.

5. Conclusions

In this work, the power flow phenomena of single-phase elements were investigated and modeled as IPPF. The analysis showed that IPPF can be decomposed into three components: the fundamental power of the single-phase elements which is drawn equally from each terminal, the IPPR caused by voltage unbalance, and IPPR caused by the presence of the opposite power quantity. The last component is the key that enables IPPF controls for (1) line-to-line single-phase elements and (2) line-to-neutral single-phase elements that are connected at the wye side of delta-wye transformers.

Based on the developed model, the control for line-to-line SVCs and power electronic interfaces was proposed for achieving a desired IPPR or IPPF. For SVCs, the determination of the appropriate reactive power and reactance for achieving the desired power routing was proposed. For the power electronic interfaces, two operation modes along with the hierarchical controls were proposed. In the first mode, the total power of the power electronic interfaces is controlled. In the second mode, ancillary functions including precise power injection and interphase power routing control were developed.

The developed models and control effectiveness were verified and demonstrated through two simulations. In the multimicrogrid system simulation, IPPF control was implemented to improve system operations and flexibility by directing the active power from a phase of a microgrid to the desired device in another phase of another microgrid. The main active power operations of the controlled devices are not interrupted while the devices are utilized for IPPF control. Furthermore, the control burden can be shared among multiple participants for distributed control. In the second simulation, a simple IPPF control was applied to a utility distribution circuit to balance the active power at the feeder head. Results showed reductions in both system voltage unbalance and losses.

Future work should explore improved coordination for IPPF controls to further enhance the operation and resiliency of distribution systems and microgrids. In addition to the line-to-line single-phase elements, other devices should be investigated for potential IPPF control utilization. The performance of the future IPPF controls should be improved with more complex dispatch schemes. Reactive power routing and limitations of IPPF devices such as energy and power availability should be considered.

Author Contributions: Conceptualization, P.S., S.S.; Validation, P.S., V.C.C.; Writing—original draft, P.S.; Writing—review and editing, N.G.B., S.S., V.C.C. All authors have read and agreed to the published version of the manuscript.

Funding: This work was supported in part by Sao Paulo Research Foundation (FAPESP) grants 2017/10476-3, 2019/20186-8 and 2016/08645-9.

Institutional Review Board Statement: Not applicable.

Informed Consent Statement: Not applicable.

Data Availability Statement: No new data were created or analyzed in this study. Data sharing is not applicable to this article.

Conflicts of Interest: The authors declare no conflict of interest.

Abbreviations

The following abbreviations are used in this manuscript:

DER	Distributed energy resource
DSRF	Double synchronous reference frame
DSTATCOM	Distribution static compensator
ES	Energy storage
FACT	Flexible alternating current transmission system
IPPF	Interphase power flow
IPPR	Interphase power routing
MC	Microgrid
P	Active power
PLL	Phase lock loop
PV	Photovoltaic
Q	Reactive power
S	Complex power
SOGI-QSG	Second-order-general integrator quadrature signal generator
SVC	Static var compensator

References

1. Chembe, D. Reduction of power losses using phase load balancing method in power networks. In Proceedings of the World Congress on Engineering and Computer Science (WCECS), San Francisco, CA, USA, 20–22 October 2009; pp. 492–497.
2. Sahito, A.; Memon, Z.; Shaikh, P.; Ahmed Rajper, A.; Memon, S. Unbalanced loading; An overlooked contributor to power losses in HESCO. *Sindh Univ. Res. J. Sci. Ser.* **2016**, *47*, 779–782.
3. Ghosh, A.; Joshi, A. A new approach to load balancing and power factor correction in power distribution system. *IEEE Trans. Power Deliv.* **2000**, *15*, 417–422. [CrossRef]
4. Czarnecki, L.S.; Haley, P.M. Unbalanced power in four-wire systems and its reactive compensation. *IEEE Trans. Power Deliv.* **2015**, *30*, 53–63. [CrossRef]
5. Quintela, F.; Arévalo, J.; Redondo, R.; Melchor, N. Four-wire three-phase load balancing with Static VAr Compensators. *Int. J. Electr. Power Energy Syst.* **2011**, *33*, 562–568. [CrossRef]

6. Jiang, I.H.; Nam, G.; Chang, H.; Nassif, S.R.; Hayes, J. Smart grid load balancing techniques via simultaneous switch/tie-line/wire configurations. In Proceedings of the IEEE/ACM International Conference on Computer-Aided Design (ICCAD), San Jose, CA, USA, 3–6 November 2014; pp. 382–388.

7. Haq, S.U.; Arif, B.; Khan, A.; Ahmed, J. Automatic three phase load balancing system by ssing fast switching relay in three phase distribution system. In Proceedings of the 1st International Conference on Power, Energy and Smart Grid (ICPESG), Mirpur, Pakistan, 12–13 April 2018; pp. 1–6.

8. Chang, T.H.; Lee, T.E.; Lin, C.H. Distribution network reconfiguration for load balancing with a colored petri net algorithm. In Proceedings of the 2017 ICASI, Sapporo, Japan, 13–17 May 2017; pp. 1040–1043.

9. Mayordomo, J.G.; Izzeddine, M.; Asensi, R. Load and voltage balancing in harmonic power flows by means of Static VAr Compensators. *IEEE Trans. Power Deliv.* **2002**, *17*, 761–769. [CrossRef]

10. Mishra, M.K.; Ghosh, A.; Joshi, A.; Suryawanshi, H.M. A novel method of load compensation under unbalanced and distorted voltages. *IEEE Trans Power Deliv.* **2007**, *22*, 288–295. [CrossRef]

11. Singh, B.; Solanki, J. A comparison of control algorithms for DSTATCOM. *IEEE Trans. Ind. Electron.* **2009**, *56*, 2738–2745. [CrossRef]

12. Arya, S.R.; Singh, B.; Niwas, R.; Chandra, A.; Al-Haddad, K. Power quality enhancement using DSTATCOM in distributed power generation system. *IEEE Trans. Ind. Appl.* **2016**, *52*, 5203–5212. [CrossRef]

13. Chang, W.; Liao, C.; Wang, P. Unbalanced load dompensation in three-phase power system with a current-regulared dstatcom based on multilevel converter. *J. Mar. Sci. Technol.* **2016**, *24*, 484–492.

14. Tenti, P.; Trombetti, D.; Tedeschi, E.; Mattavelli, P. Compensation of load unbalance, reactive power and harmonic distortion by cooperative operation of distributed compensators. In Proceedings of the 2009 13th European Conference on Power Electronics and Applications, Barcelona, Spain, 8–10 September 2009; pp. 1–10.

15. Mousazadeh Mousavi, S.Y.; Jalilian, A.; Savaghebi, M.; Guerrero, J.M. Flexible compensation of voltage and current unbalance and harmonics in microgrids. *Energies* **2017**, *10*, 1568. [CrossRef]

16. El-Naggar, A.; Erlich, I. Control approach of three-phase grid connected PV inverters for voltage unbalance mitigation in low-voltage distribution grids. *IET Renew. Power Gener.* **2016**, *10*, 1577–1586. [CrossRef]

17. Hong, T.; de León, F. Controlling non-synchronous microgrids for load balancing of radial distribution systems. *IEEE Trans. Smart Grid* **2017**, *8*, 2608–2616. [CrossRef]

18. Tang, F.; Zhou, J.; Xin, Z.; Huang, S.; Loh, P.C. An improved three-phase voltage source converter with high-performance operation under unbalanced conditions. *IEEE Access* **2018**, *6*, 15908–15918. [CrossRef]

19. Ciobotaru, M.; Teodorescu, R.; Blaabjerg, F. A new single-phase PLL structure based on second order generalized integrator. In Proceedings of the 2006 37th IEEE Power Electronics Specialists Conference, Jeju, Korea, 18 June 2006; pp. 1–6.

20. Reno, M.J.; Coogan, K. Grid Integrated Distributed PV (GridPV). Available online: https://www.osti.gov/servlets/purl/1165229 (accessed on 10 January 2015).

Article

Optimal Operation of Solar Powered Electric Vehicle Parking Lots Considering Different Photovoltaic Technologies

Mahsa Z. Farahmand [1], Sara Javadi [2], Sayyed Muhammad Bagher Sadati [3], Hannu Laaksonen [2] and Miadreza Shafie-khah [2,*]

[1] Department of Electrical Engineering, K. N. Toosi University of Technology, Tehran 19697-64499, Iran; mahsafarahmand23@gmail.com

[2] School of Technology and Innovations, University of Vaasa, FI-65200 Vaasa, Finland; sara.javadi@uwasa.fi (S.J.); hannu.laaksonen@univaasa.fi (H.L.)

[3] National Iranian Oil Company (NIOC), Iranian Central Oil Fields Company (ICOFC), West Oil and Gas Production Company (WOGPC), Kermanshah 67146-77745, Iran; bagher_sadati@yahoo.com

* Correspondence: mshafiek@univaasa.fi

Abstract: The performance of electric vehicles and their abilities to reduce fossil fuel consumption and air pollution on one hand and the use of photovoltaic (PV) panels in energy production, on the other hand, has encouraged parking lot operators (PLO) to participate in the energy market to gain more profit. However, there are several challenges such as different technologies of photovoltaic panels that make the problem complex in terms of installation cost, efficiency, available output power and dependency on environmental temperature. Therefore, the aim of this study is to maximize the PLO's operational profit under the time of use energy pricing scheme by investigating the effects of different PV panel technologies on energy production and finding the best strategy for optimal operation of PVs and electric vehicle (EV) parking lots which is achieved by means of market and EV owners' interaction. For the accurate investigation, four different PV panel technologies are considered in different seasons, with significant differences in daylight times, in Helsinki, Finland.

Keywords: solar-powered electric vehicle parking lots; different PV technologies; PLO's profit; uncertainties

Citation: Farahmand, M.Z.; Javadi, S.; Sadati, S.M.B.; Laaksonen, H.; Shafie-khah, M. Optimal Operation of Solar Powered Electric Vehicle Parking Lots Considering Different Photovoltaic Technologies. *Clean Technol.* **2021**, *3*, 503–518. https://doi.org/10.3390/cleantechnol3020030

Academic Editor: Hamidreza Nazaripouya

Received: 26 March 2021
Accepted: 1 June 2021
Published: 16 June 2021

Publisher's Note: MDPI stays neutral with regard to jurisdictional claims in published maps and institutional affiliations.

1. Introduction

In the last few years using renewable energy resources and the new generation of transportation systems has increased enormously due to the issues like lack of fossil fuel resources, carbon emission and environmental issues in a way that their role in the future of power systems is very important and undeniable. Thus progression pace and finding potentials in resources are crucial points for the energy issue. Among all other renewable resources, solar energy is the most plentiful resource on the planet earth and there is a high possibility to utilize its sources by means of photovoltaic panels. On the other hand, the arrival of electric vehicles in the transportation system has made the market of this product very competitive due to the many environmental advantages they have and their role in transportation in the near future. However, using these sources all together brings forth the necessity of facing challenges like stability and reliability of the grid, which requires fundamental preparations such as control managements, cooperation agencies, etc. On the other hand, using PV and EV simultaneously can be very useful and beneficial for the grid in the field of energy supplement and economic opportunities and this exploitation can be taken into action through electric vehicles parking lots. Since electric vehicles are mostly parked in the parking lots during the day, employing rooftop photovoltaic for the parking lots makes great sense. However, there are several challenges such as different technologies of photovoltaic panels that make the problem complex in terms of installation cost, efficiency, available output power and dependency on environmental temperature.

Since changing weather conditions can affect the output current and voltage, the response of the PV system to these changes needs to be characterized [1,2]. In order to estimate the amount of energy a solar panel can generate in a lifetime, weather data and solar irradiation information are utilized. Most solar panels do not operate in an ideal condition, because the weather is always changing. By knowing the reaction of solar panels to different weather conditions, it is possible to improve their efficiency in non-optimal conditions [3]. In some situations, for example, in hot climates, a cooling system is needed to keep the panels in certain temperatures. In addition to the temperature situation, the PV panel material is also important to predict the output power, because the efficiency of different materials has different levels of dependency on temperature. The temperature coefficient describes the material temperature dependency [4].

Due to the fact that EV parking lots are magnificently useful for EVs, they can be combined with PV as a source of independent energy supply that can eventually decrease environmental damage such as greenhouse gas emission and even bring more benefit to suppliers and consumers. Besides, this breeds a condition in which there are financial and technical advantages in EV parked time and also makes it easier to interact with the supply and demand market.

According to the mentioned points, the goal of this study is to maximize the benefit for the parking lot operators by finding the best strategy for the optimal operation of PVs and EV parking lots which is achieved by means of market and EV owner interaction, considering the fact that the distribution network, uncertain behavior of different PV technologies and EV owners can have an impact on the behavior of PEV parking lots.

1.1. Literature Review

EVs reaching the number of 5 million in 2018 is a clear fact indicating that the future of the automobile industry belongs to EVs. It is also expected that this number will reach 250 million by 2030. According to the EV 30@30 scenario, 44 million EVs will be sold each year [5]. The enormous growth in EV production logically brings forth the need of building charging stations (CS) and parking lots (PLs). According to [5], as of 2018, 5.2 million charging stations are available and almost all of them (about 90%) are private.

Seeking the highest economic efficiency, PL operators (PLO) must try to satisfy EV owners (EVO) through some strategies. EVs parked time is the key to the goal, and it brings the opportunity for EVOs to cooperate with parking lot operators to sell their EV's discharging power to distribution network operators or even other parking lot operators. To satisfy EVO's, parking lot operators should dedicate a reasonable percentage of the benefit to them by means of a written contract so that they are motivated to sell even more of this discharging energy to parking lot operators. This however brings out the necessity of a smart energy management system (EMS) in the PL so that maximum profit is gained by an optimized charge/discharge program. Additionally, using renewable energy resources is a way to increase profit. According to distribution system operator's (DSO) tariff, some of the studies on EV charge/discharge programs in solar-powered and non-solar-powered EVPL's will be reviewed accordingly.

In Ref. [6], a mathematical model for estimating the electricity capacity of a PV parking lot is described, and new formulas are proposed to investigate the effect of batteries and inverters on the power demand during battery charging and discharging. The results show that the use of PV panels in the parking lot can reduce the load of the distribution grid by reducing the effective load during peak charging.

In Ref. [7], to reach the highest profit for an EVPL using solar panels and distributed generators a self-scheduling model considering spinning reserve is investigated. Ref. [8] works on large-scale wind integration and operational flexibility of parking lots by introducing a two-stage stochastic model. Due to the lack of enough flexible resources, a lot of wind energy is wasted. The use of parking lots, not only reduces the cost of operation but also using the potential flexibility and participating in the energy and reserve markets can reduce wind spillage. Ref. [9] tries to reduce EV charging costs to the lowest level by producing a Convexfiel model (the model is obtained from the conventional model by using convex relaxation techniques) in which EVs uncertainty and V2G ability are considered. In Ref. [10], to reduce the daily cost in an EVPL that uses solar energy to the lowest level, for EV charging, mixed-integer linear programming (MILP) is proposed.

Ref. [11] worked on the potential of solar energy for charging EVs and reducing the payback time. Hence, a Genetic algorithm is used to increase the production of solar energy, energy storage and smart charging is utilized to investigate different charging methods. By applying the proposed model, the payback time is reduced from 14 years to 7 years. Ref. [12] is about EVPL operational scheduling in the energy and reserve market. In this study, a bi-level model is estimated in which the upper level's goal is to bring down the operation cost to the lowest amount and the lower level's goal is to lessen PL cost. Ref. [13] considers maximizing parking lot operators' profit by controlling EV charging and offers a dynamic charging program in order to achieve the goal.

Ref. [14] worked on a solar-powered charging station using a fixed battery, providing an algorithm consisting of four stages to minimize operation cost on EV charge/discharge by most optimization and customer satisfaction. In [15], as in [16], a multi-objective model is proposed with the aim of EVPL cost minimization considering climate effects. This model shows that an appropriate charge/discharge program for EVs can result in less total emission and operation costs. In Ref. [17] the goal is to decrease electricity tariff through optimized EV power charge/discharge in a solar electric vehicle parking lot (EVPL) (in an EVPL using SE). In [18], an EV charge model based on EVO satisfaction, cost minimization in an integrated EVPL using solar energy and an energy storage system is proposed.

In Ref. [19], considering environmental and economic targets, a model giving a schedule for EV charge/discharge is proposed with two main objects consisting of emission reduction and cost operation minimization. In Ref. [20] an energy management strategy is suggested for a solar energy EVPL to analyze its effects on loss reduction and power consumption of distribution network. In Ref. [21], regarding the reduction of charging costs optimized with a photovoltaic system in an EVPL, two-stage stochastic mixed-integer linear programming (MILP) is proposed. A stochastic optimized energy management program considering both parking lots' operators and EVO's benefits is offered in Ref. [22]. In order to lower the cost for various grid purposes and regarding EV's uncertainty, an optimized program for EVs is designed in Ref. [23]. In Ref. [24] two cases are studied. The first one refers to a risk-based model that analyzes the efficiency of EVPLs using hydrogen storage systems, solar energy, etc. The second is dedicated to the charge/discharge program for EVs in risk-averse and risk-neutral performance. Ref. [25] refers to charge/discharge energy trading with DSO and also cooperation between parking lot operators and EVOs to offer a program to EV aggregators considering the highest benefit for both sides.

Ref. [26] studied the relationship between the amount of output energy and variation of temperature. In order to show the influence of temperature on photovoltaic systems, two models were used. Model A ignored temperature and Model B considered it. These two models were carried out for 236 cities in America. In the Northeast and the Midwest regions, Model B power outputs were higher in comparison with Model A (16–20%), from November to February, whereas there was a reduction from May to August (−4%). Instead, in the South and Southwest of America, power outputs reduced significantly from May to August (−12–15%), whereas there was a slight increase from December to February (5%). In Ref. [7], the effect of temperature on the performance of different photovoltaic technologies was evaluated in Amman, Jordan. Three photovoltaic systems (Poly-crystalline, Mono-crystalline and Thin-film) with the same design parameters were chosen. It was shown that the temperature has less effect on the thin-film solar cells. Ref. [27] evaluated the temperature coefficient for some different types of commercially accessible solar panels. The tests were done at the PV test facility of the Solar Energy Centre, New Delhi, India. The panels were chosen randomly from different manufactures. The study showed that the temperature coefficient for the monocrystalline silicon module is higher than the other types.

1.2. Contributions

According to the literature, several studies have been conducted on energy exchange between PL and DS and the performance of different PV technologies has been studied separately in which the effect of different PV technologies and their performance on the possibility of energy exchange and profit of the parking lot operators are not taken into account.

The goal of this study is to maximize the benefit to parking lot operators in the energy market between DSO and EVOs in the rooftop PV parking lot with different PV technologies. To meet this goal, different scenarios are considered in detail to evaluate and analyze operations under various circumstances. Also, the model considers all the impacts of the behavior of different PV technologies.

To the best of the authors' knowledge, the impact of the PV technologies on the optimal operation of rooftop PV parking lots has not been reported in the literature, which is the main contribution of the paper. In order to investigate the impact of PV technologies and their level of sensitivity to temperature and solar radiation, different PV technologies such as Monocrystalline silicon, Polycrystalline silicon, Amorphous silicon and Cadmium Telluride based solar modules were considered in four different months. Another applicable contribution of the paper is that due to very different daylight times in Finland in different months of a year, there are huge changes on the operation of rooftop PV parking lots.

In Section 2, the mathematical model of a PEV parking lot is formulated. The case study is described in Section 3. In Section 4, the results of the study are analyzed. The conclusion is presented in Section 5.

2. Problem Formulation

The objective function of this study is to maximize the profit for parking lot operators by using four different PV technologies including Monocrystalline silicon, Polycrystalline silicon, Amorphous silicon and Cadmium Telluride to evaluate which one is the best. Considering that the uncertain behaviors of solar irradiation and EVO have a direct effect on the profit of the parking lot operators, the influence of these factors was investigated in order to get the optimum result. Hence, by using the Beta function, the uncertainty of solar radiation was modeled and the truncated Gaussian distribution, that is, Normal PDF was applied for the other uncertainties that all the required equations were taken from [23]. Also, the complexity of the calculations caused a reduction in the scenarios. Therefore, in this study, by using a scenario reduction method, similar scenarios were deleted.

2.1. Objective Function

The objective function, presented in (1), represents different income terms that maximize the profit from the parking lot operator's point of view. These terms show cooperation between the DSO, EVOs and parking lot operators. Therefore, the objective function of parking lot operators includes the revenue such as selling the power generated by PV rooftops to EVOs and DSO and selling EVs discharging energy to DSO. Also, purchasing energy from DSO and depreciation of the battery because of selling energy to DSO are the cost terms of the objective function. To encourage EVOs, some money is given to them for several discharges in one day and they can also receive a portion of the revenue gained from the sale of energy to the DSO. Also, based on the energy price, parking lot operators can sell the power generated by PV to DSO and EVOs.

$$
\begin{aligned}
\text{Profit}_{PL} = \sum_{n}\sum_{h} & \left(\left(P_{n,h}^{ch-Solar}\text{Pr}_{h}^{ch} \right) + \left(P_{h}^{Solar}\text{Pr}_{h}^{dch} \right) \right)\Delta h \\
+\sum_{w}\rho_{w}\sum_{n}\sum_{h} & \left(\begin{array}{l} \left(P_{n,h,w}^{ch-Grid}\text{Pr}_{h}^{ch} \right) + \left((1-\alpha)P_{n,h,w}^{dch-Grid}\text{Pr}_{h}^{dch} \right) \\ -\left(P_{n,h,w}^{ch-Grid}\text{Pr}_{h}^{Grid} \right) - \left(P_{n,h,w}^{dch-Grid}C^{cd} \right) \end{array} \right)\Delta h
\end{aligned}
\tag{1}
$$

In (1), $P^{ch\text{-}Solar}$ is Charging power of each EV from solar energy's output, P^{Solar} is Solar energy's output of PL to DSO, $P^{ch\text{-}Grid}$ is Charging power of each EV from DSO and $P^{dch\text{-}Grid}$ is Discharging power of PL to DSO.

2.2. Constraints

Arrival/departure times of EVs to/from the *PL* and their duration of presence in *PL* are presented in (2) to (4).

$$SOE_{n,h,w} = SOE^{arv}_{n,h^{arv},w} - \left(\frac{P^{dch}_{n,h,w}}{\eta^{dch}}\right) + \left(P^{ch}_{n,h,w}\right)\eta^{ch} : \forall PL, v, h = h^{arv}, w \qquad (2)$$

$$SOE_{n,h,w} = SOE_{n,h-1,w} - \left(\frac{P^{dch}_{n,h,w}}{\eta^{dch}}\right) + \left(P^{ch}_{n,h,w}\right)\eta^{ch} : \forall PL, v, h \succ h^{arv}, w \qquad (3)$$

$$SOE_{n,h,w} \geq SOE^{dep}_{n,h^{dep}} : \forall n, h = h^{dep}, w \qquad (4)$$

It is impossible for a battery to charge and discharge, simultaneously:

$$X^{ch}_{n,h,w} + X^{dch}_{n,h,w} \leq 1; \forall n, h, w \qquad (5)$$

According to the rate of charge and discharge of EV batteries, the value of *SOE* is shown in (6).

$$SOE^{min}_{n,h,w} \leq SOE_{n,h,w} \leq SOE^{max}_{n,h,w} : \forall n, h, w \qquad (6)$$

EVs can purchase their required energy at the mid-peak and off-peak periods from DSO and the rate of charging/discharging is restricted between zero and nominal rate. Also, the EV charging/discharging time is not the same. The mentioned constraints are shown in (7) and (8).

$$0 \leq P^{ch}_{PL,v,h,w} = P^{ch\text{-}Grid}_{n,h,w} + P^{ch\text{-}Solar}_{n,h} \leq X^{ch}_{n,h,w} \times R_{ch,max} : \forall n, h, w \qquad (7)$$

$$0 \leq P^{dch}_{n,h,w} = P^{dch\text{-}Grid}_{n,h,w} \leq X^{dch}_{n,h,w} \times R_{dch,max} : \forall n, h, w \qquad (8)$$

Based on (9), the parking lot operator sells energy to DSO after 24-h periods. Equation (10) also guarantees that the amount of EV charging through PV generation and the amount of PV generation sold to the DSO is equal to the output of the PV panel in PL. Each EV's charging and discharging power is restricted to four times the nominal rate and is shown in (11) and (12).

$$0 \leq P^{Solar}_{h} \leq P_{solar,h} : \forall solar, h \qquad (9)$$

$$P^{Solar}_{h} + \sum_{n=1}^{100} P^{ch\text{-}Solar}_{n,h} = P_{solar,h} : \forall solar, h \qquad (10)$$

$$\sum_{h=1}^{24} P^{ch\text{-}Grid}_{n,h,w} \leq 4 \times R_{ch,max} : \forall n, w \qquad (11)$$

$$\sum_{h=1}^{24} P^{dch\text{-}Grid}_{n,h,w} \leq 4 \times R_{dch,max} : \forall n, w \qquad (12)$$

2.3. Equations for PV Generation

Considering the change of solar radiation and temperature under outdoor conditions, the short-circuit current (ISC) and open-circuit voltage (VOC) of silicon-based solar cells are expressed as follows [28]:

$$I_{SC} = \left(\frac{G}{G_{STC}}\right) I_{SC\_STC}[1 + a(T_C - T_{STC})] \qquad (13)$$

$$V_{OC} = V_{OC\_STC}[1 + \beta(G_{STC})(T_C - T_{STC})][1 + \delta(T)\ln\left(\frac{G}{G_{STC}}\right)] \tag{14}$$

where G is the solar radiation of PV solar cell under operating conditions ($\frac{W}{m^2}$); G_{STC} is the solar radiation under standard test conditions (($\frac{W}{m^2}$) and G_{STC} = 1000 $\frac{W}{m^2}$); T_{STC} is the temperature under standard conditions (T_{STC} = 25 °C); a is the temperature coefficient of short-circuit current, usually provided by the manufacturer; I_{SC_STC} is the short-circuit current of solar cell under standard test conditions; V_{OC_STC} is the open-circuit voltage of solar cell under the standard test conditions; β is the temperature coefficient of open-circuit voltage; and δ is the correction factor of solar radiation. As a result, as the temperature increases, the open-circuit voltage becomes smaller and the short-circuit current becomes larger, which leads to a decrease in efficiency. The specific relationship can be expressed as follows:

$$\eta = \eta_{ref}[1 - \beta_{ref}(T_C - T_{STC})] + \gamma\log_{10} G \tag{15}$$

where η_{ref} is the power generation efficiency of solar cells under standard test conditions; γ is the solar radiation coefficient; and β_{ref} is the temperature coefficient (K−1) at reference conditions. In engineering applications, the relationship between the output power of PV and the temperature can be described as follows [29]:

$$P_{TC} = \eta_{ref}[1 - \beta_{ref}(T_C - T_{STC})]GA \tag{16}$$

where, A is the surface area of the PV module.

3. Case Study

In order to evaluate the different PV technologies a PEV parking lot with a capacity of 80 EVs was considered. Figure 1 illustrates the interaction between all three components which result from energy exchange and individual contracts with EVOs. In this evaluation, three different scenarios were studied. Scenario I represents the base case with no PV panels in four different months including February, May, August and November. In scenario II, the PV rooftop (area of panels is approximately 558 m^2) with four different PV technologies including Monocrystalline silicon, Polycrystalline silicon, Amorphous silicon and Cadmium Telluride was analyzed in two cold periods (February and November). Similarly, in scenario III, a PV rooftop with four different PV technologies was investigated in two warm periods (May and August).

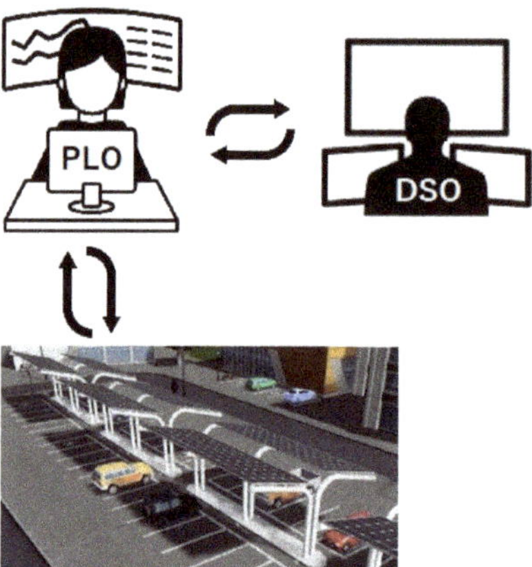

Figure 1. Interaction between PLO, DSO and PEV.

Since the uncertainties of solar radiation have a direct effect on PV power generation, 10 scenarios were considered for different months and seasons (February, May, August and November). To calculate the output power of PV, the real solar radiation data of Helsinki, Finland was used [30]. Figure 2 shows the expected value of PV generation for each technology in each month.

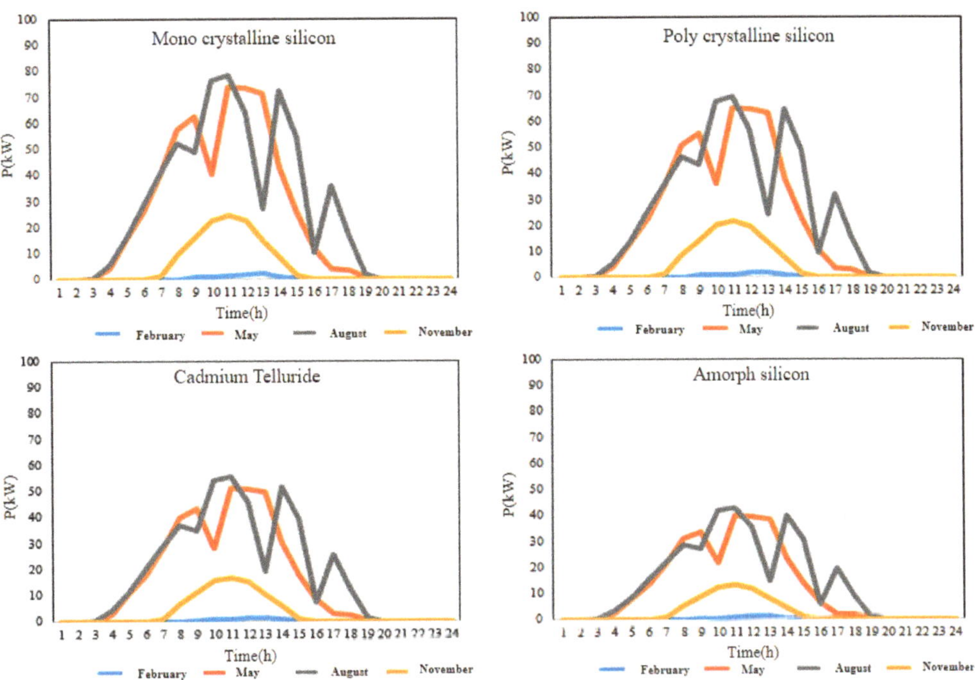

Figure 2. Expected value of output power of PV on each month for each technology.

The main specifications of each technology are represented in Table 1.

Table 1. The specifications of each technology.

Cell Type	Structure	Efficiency [1–3]	Thermal Coefficient (C^{-1})
Monocrystalline silicon	• manufactured from pure semiconducting material with no defects or impurities in the silicon crystalline structure • production procedure is complicated • higher-priced than some other technologies	17–22%	0.0044
Polycrystalline silicon	• manufacturing process is simpler than the monocrystalline ones • more cost-effective • more defects in the crystalline structures	15–17%	0.0038
Thin film solar panels: Amorphous silicon Cadmium Telluride	• completely different from crystalline solar panels • lightweight and, in some cases, flexible	10–13%	0.0023 0.0017

It is worth mentioning that the efficiency considered in the calculation for Monocrystalline silicon, Polycrystalline silicon, Amorphous silicon and Cadmium Telluride is 18%, 16%, 10% and 13% respectively. Also, the EVs' probability distribution and their specification are presented in Tables 2 and 3 respectively. The rate of charge and discharge of each EV is up to 40 kWh, and they were charged and discharged four times with maximum rates (10 kWh).

Table 2. The modified probability distribution of EVs [31].

	Mean	Standard Deviation	Minimum	Maximum
Initial SOC (%)	54.85	8.92	40	70
The time of Arrival (h)	8	3	5	23
The time of Departure (h)	16	3	6	24
Cap (kW)	62.79	28.60	18	95

Table 3. EVs specification.

Capacity of Battery	32 kWh	Efficiency of Charge	90%
The rate of Charging/discharging	10 kW	Efficiency of Discharge	95%
SOE min	4.8 kWh	Desired SOE	28.8 kWh
SOE max	28.8 kWh	Ccd	20 €/MWh

The price of energy data was drawn from the Finnish electricity market [32,33]. The price of energy exchange is presented in Table 4. Also, the price of selling energy to DSO is shown in Figure 3.

Table 4. The price of energy exchanged between DSO and PLO (cent/kWh).

Price/Time	Off-Peak Periods (01:00–06:00)	Mid-Peak Periods (07:00–11:00, 18:00–19:00)	On-Peak Periods (12:00–17:00)
Energy purchased from DSO by the PLOs	4.19	4.19	4.19
Energy sold to EVO by the PLO	4.67	5.53	5.53

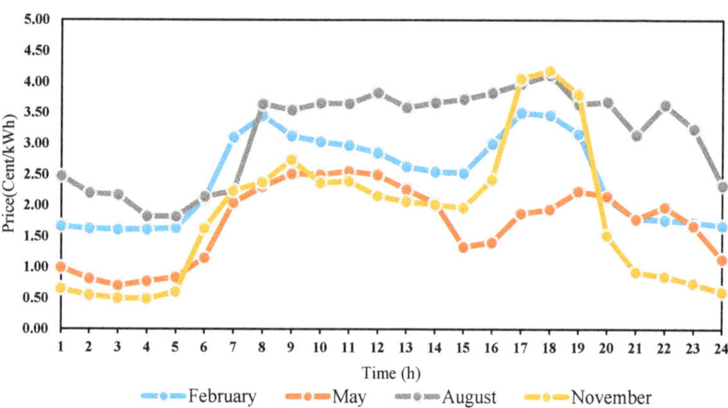

Figure 3. Price of selling energy to DSO.

4. Results and Discussion

4.1. PLO's Profit without PV Generation

In this scenario, the energy exchange is between DSO and EVOs and each EV makes revenue for the PV parking lot operators by selling energy to DSO. On the other hand, based on the EVO's contract, each EV can gain benefit from battery depreciation cost and half of the selling energies' revenue to DSO. Figure 4 shows the required energy for charging EVs every hour of each month. Based on Figure 4, the entrance time of EVs to PL starts at 8:00 approximately. Between 8:00 and 20:00 more EVs enter the PL which increases the energy exchange between PL and DSO.

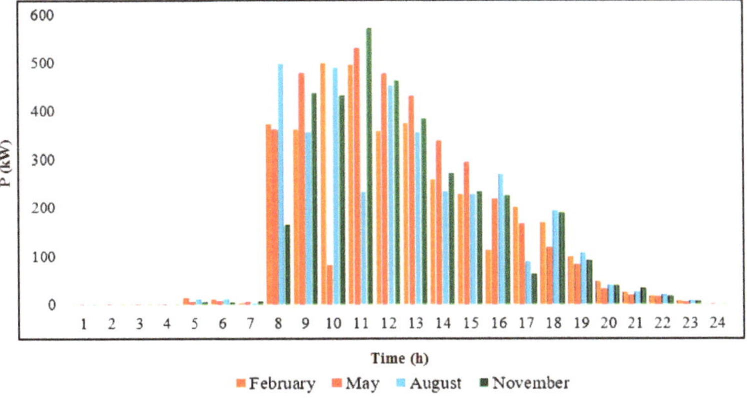

Figure 4. Required energy for charging EVs each month.

The charging/discharging energy of all EVs in four months and the profit of the parking lot operator are presented in Table 5. According to Table 5, in August, the energy interaction between the parking lot operator and DSO is more than that in the other months which brings more profit to the parking lot operator.

Table 5. Charging/discharging energy of EVs and PLO's profit.

Month	February	May	August	November
Charging (kWh)	3636.94	3636.94	3608.85	3620.89
Discharging (kWh)	1123.22	1128.01	1311.24	1285.6
PLO's profit (€)	44.70	40.25	47.33	42.65

4.2. PLO's Profit with PV Generation

In this scenario, for evaluating the effect of the PV panel technology on parking lot operator profit, four different technologies including Monocrystalline silicon, Polycrystalline silicon, Amorphous silicon and Cadmium Telluride were considered. Due to the uncertainty of solar energy, four months were considered for a more accurate evaluation. In this regard, trading energy between PL and DSO is shown in Table 6.

Table 6. Energy exchange between PL and DSO (kWh).

	February		May		August		November	
	Pch from DSO	Pch to DSO	Pch from DSO	Pch to DSO	Pch from DSO	Pch to DSO	Pch from DSO	Pch to DSO
Without PV	3636.94	1123.22	3636.94	1128.01	3608.85	1311.24	3620.89	1285.60
Mono crystalline	3628.06	1118.80	3625.13	1267.40	3598.09	1340.39	3600.57	1283.05
Poly crystalline	3628.13	1118.44	3627.21	1246.91	3598.1	1313.35	3600.52	1279.27
Amorphous silicon	3628.84	1117.95	3632.94	1192.80	3598.9	1239.24	3602.47	1277
Cadmium Telluride	3628.44	1118.09	3627.70	1212.60	3595.93	1272.83	3602.98	1281.32

For better understanding the effect of PV rooftops on EV parking lots and parking lot operator profit, Figures 5–8 are given. These figures show how much energy in a day is supplied by DSO and PV panels. Before 7:00, the PV generation is at its lowest level and EVs can only be charged by purchasing energy from the DSO. Between 8:00 to 14:00, due to the production of more energy by the PV rooftops, EVs are charged by the DSO and solar energy. During this time the amount of energy purchased from the DSO is reduced. The amount of reduction in energy purchases from the DSO depends on the power generated by PV and the decision to sell solar energy based on the price of energy. After 14:00 by reducing the energy of PV, the DSO can charge EVs. As can be seen in February and November, the highest amount of energy is supplied through the DSO because in these two months, the solar radiation is very low, and the PV panels are not able to produce energy. While the energy generated by PV panels with the four PV technologies in May and August is quite clear.

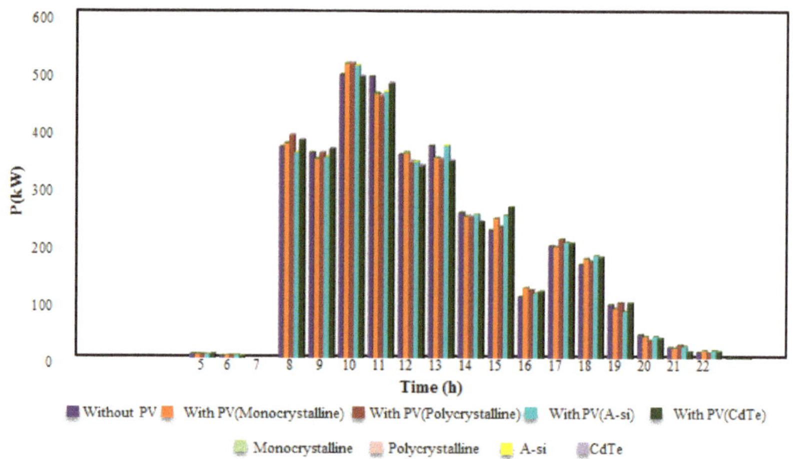

Figure 5. Power supplied by DSO and PV panel in February.

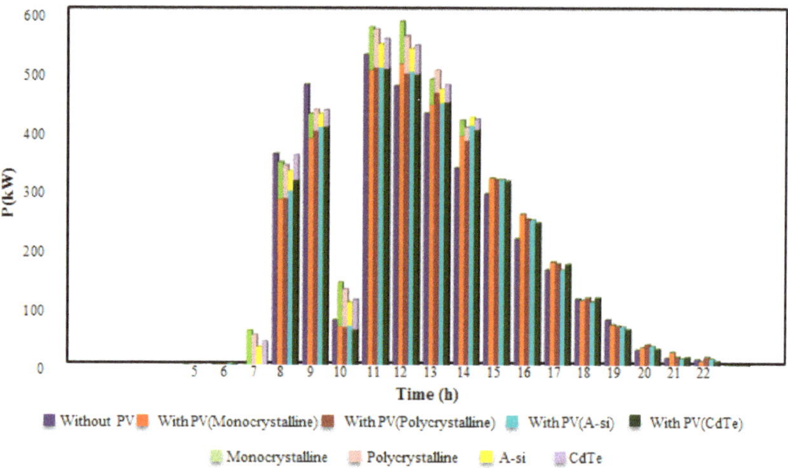

Figure 6. Power supplied by DSO and PV panel in May.

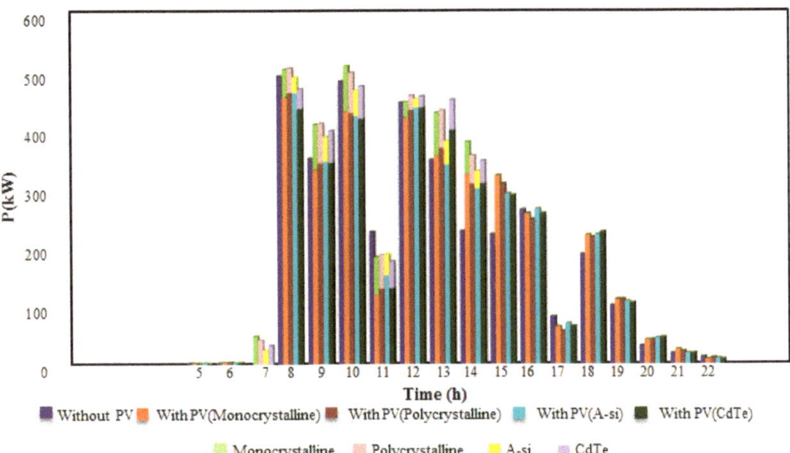

Figure 7. Power supplied by DSO and PV panel in August.

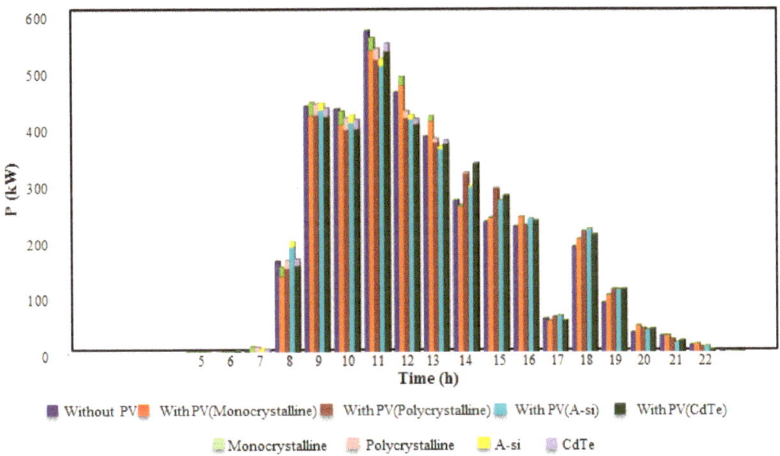

Figure 8. Power supplied by DSO and PV panel in November.

Discharging energy of PL to DSO is presented in Figures 9–12. According to these Figures and Figure 3, the energy exchange between PL and DSO depends on PV generation, number of EVs and the price of selling the energy to DSO. In February at 7:00 and 16:00 and in November at 8:00 and 17:00, the number of EVs in the PL is less and the price of selling energy is at the highest amount, therefore the energy sold to DSO by the parking lot operator is a significant amount. In May at 8:00 and 10:00 and in August at 11:00 and 17:00, by reducing the number of EVs, increasing PV generation and because of the selling energy price, a lot of energy is traded between the parking lot operator and DSO.

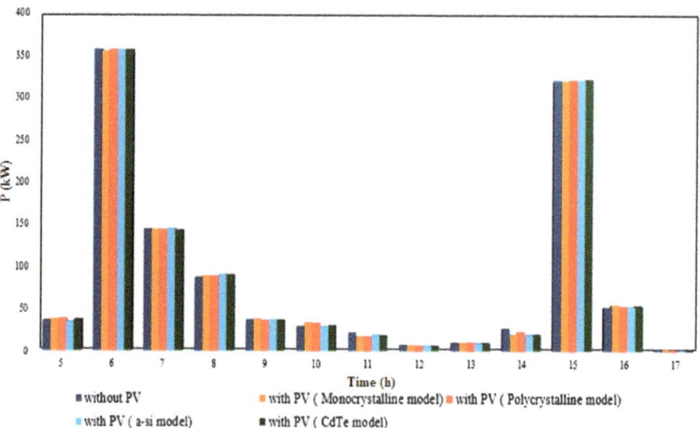

Figure 9. Discharging energy from PL to DSO in February.

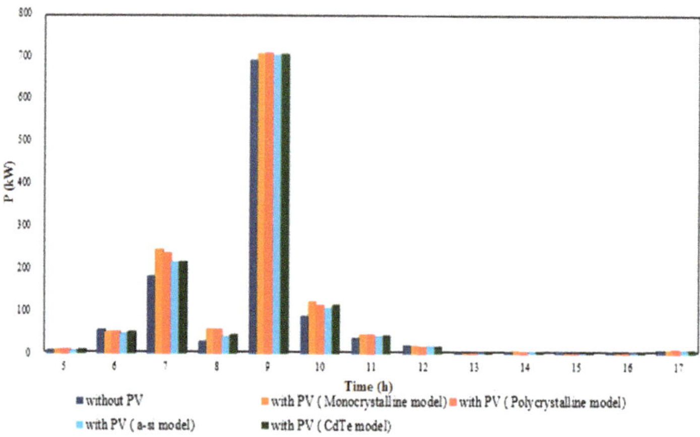

Figure 10. Discharging energy from PL to DSO in May.

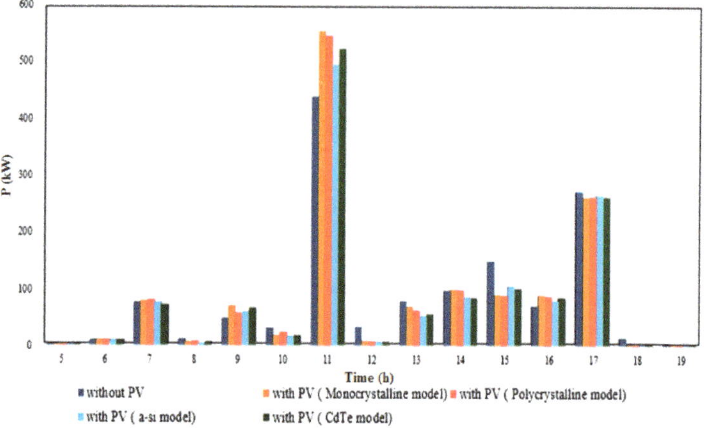

Figure 11. Discharging energy from PL to DSO in August.

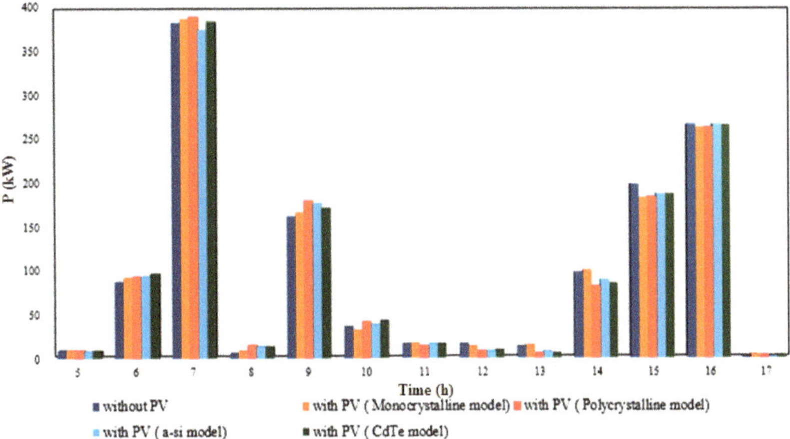

Figure 12. Discharging energy from PL to DSO in November.

As it can be seen, installing PV rooftops on EV parking lots reduces the energy purchased from DSO to PL but the performance of the PV technology can change the amount of energy exchanged. Table 7 presents the profit of the parking lot operator in all scenarios. Based on the results, the least profitable scenario is scenario I because all the required energy is supplied by DSO due to the lack of other energy sources to support the EV charging energy. Thus the parking lot operator gains less profit.

Table 7. PLO's profit (€) by using different PV technologies.

PV Technology/Month	February	May	August	November
Without PV	44.70	40.25	47.33	42.65
Mono crystalline	45.06	63.26	72.73	48.85
Poly crystalline	45.01	60.52	69.82	48.11
Amorphous silicon	44.87	52.72	61.25	45.89
Cadmium Telluride	44.93	56.26	65.36	46.86

According to Table 7, PV generation helps parking lot operators to participate more in the energy exchange and increase their profit but because of the weather conditions, solar radiation and the type of PV technology the amount of income is different each month. In February and November, due to the low value of solar radiation, the energy produced by PV panels is less than that in May and August. However, the Monocrystalline silicon technology performance is better in all the months compared to the other technologies, because there are no defects or impurities in the silicon crystalline structure and consequently the efficiency is high. As can be observed, increasing the output power of PV allows the parking lot operator to purchase and sell energy to/from DSO at a more reasonable price and gain more profit.

The next most appropriate technology is Polycrystalline silicon. The performance of this technology is similar to Monocrystalline silicon, but there are more defects in the crystalline structures and the efficiency is slightly less than Monocrystalline cells. Thin-film solar panels like Amorphous silicon and Cadmium Telluride are completely different from crystalline solar cells and they have less efficiency. As can be seen, because of the lower performance and the lower benefits, they are not suitable for Finnish weather conditions.

Moreover, comparing profits in different months shows that the energy exchange in August is much more profitable than in other months.

5. Conclusions

The goal of this study is to maximize the benefit to parking lot operators in the energy market between DSO and EVOs for rooftop PV parking lots with different PV technologies. To analyze the impacts of different PV technologies' output power on the parking lot's profits in several weather conditions, various scenarios were considered.

Hence, by considering solar radiation uncertainty and different behaviors of EV owners, studies were performed on four different PV technologies including Monocrystalline silicon, Polycrystalline silicon, Amorphous silicon and Cadmium Telluride in February, May, August and November. Moreover, for accurate investigation, the output power curve of the PV, energy purchase and sales curve and power supplied by PV by each PV technology in 4 months were shown.

Based on the results, the parking lot's profit in February and November was lower than the other two months due to lack of solar radiation and August is the most profitable month due to suitable weather conditions, which consequently results in more PV generation. Also, a review of each technology in different months indicates that Monocrystalline silicon, every 4 months, has a better performance than other technologies, which creates a better opportunity for energy exchange and consequently a higher profit compared to the other technologies. Moreover, according to the estimated climate in the experiments, the most efficient technologies for parking lot operators are Monocrystalline silicon and Polycrystalline silicon, and the other two, Amorphous silicon and Cadmium Telluride, are not quite suitable based on the experimented circumstances. Based on the evaluation of the efficiency and performance of different technologies in Finland's weather conditions and in different scenarios, the Monocrystalline silicon technology is a more suitable option for use in parking lots. In future work, the application of these technologies in EVPLs can be compared based on the payback time and lifespan.

Author Contributions: Conceptualization, S.J., H.L., M.S.-k.; methodology, M.Z.F., S.J., S.M.B.S.; validation, S.M.B.S., M.S.-k.; investigation, H.L., M.S.-k.; writing—original draft preparation, M.Z.F., S.J.; visualization, M.Z.F., S.M.B.S.; supervision, M.S.-k. All authors have read and agreed to the published version of the manuscript.

Funding: This research received no external funding.

Institutional Review Board Statement: Not applicable.

Informed Consent Statement: Not applicable.

Data Availability Statement: Data sharing is not applicable to this article.

Acknowledgments: M. Shafie-khah and H. Laaksonen acknowledge the support by SolarX research project with financial support provided by the Business Finland, 2019–2021 (grant No. 6844/31/2018).

Conflicts of Interest: The authors declare no conflict of interest.

Nomenclature

The Main Parameters and Variables are Presented Here.

Indices

n	Index for EV number
w	Index for scenarios
h	Index for time (hour)

Parameters

C^{cd}	Cost of equipment depreciation (€/kWh)
R^{max}	Maximum charging/discharging rate (kWh)
SOE^{arv}	Initial SOE of EVs at the arrival time to PL (kWh)
SOE^{dep}	Desired SOE of EVs at the departure time from PL (kWh)
SOE^{max}	Maximum rate of SOE (kWh)

SOE^{min}	Minimum rate of SOE (kWh)
t^{arv}	Arrival time of EVs to the PL
t^{dep}	Departure time of Evs
pr^{ch}	Charging tariff of EVs (€/kWh)
pr^{Grid}	Price of purchasing energy from DSO by PL (€/kWh)
pr^{dch}	Price of selling energy to DSO by PL (€/kWh)
η^{ch}	Charging efficiency (%)
η^{dch}	Discharging efficiency (%)
ρ_ω	probability of each scenario
Variables	
$p^{ch\text{-}Grid}$	Charging power of each EV from DSO (kW)
$p^{ch\text{-}Solar}$	Charging power of each EV from solar energy's output (kW)
$p^{dch\text{-}Grid}$	Discharging power of PL to DSO (kW)
p^{Solar}	Solar energy's output of PL to DSO (kW)
SOE	EVs' state of energy (kWh)
X^{ch}	Binary variable which shows the charge status of each EV

References

1. Adeeb, J.; Farhan, A.; Al-Salaymeh, A. Temperature Effect on Performance of Different Solar Cell Technologies. *J. Ecol. Eng.* **2019**, *20*, 249–254. [CrossRef]
2. Amelia, A.R.; Irwan, Y.M.; Leow, W.Z.; Irwanto, M.; Safwati, I.; Zhafarina, M. Investigation of the Effect Temperature on Photovoltaic (PV) Panel Output Performance. *Int. J. Adv. Sci. Eng. Inf. Technol.* **2016**, *6*, 5.
3. Kawajiri, K. Effect of Temperature on PV Potential in the World. *Environ. Sci. Technol.* **2011**, *45*, 9030–9035. [CrossRef] [PubMed]
4. Kamkird, K.; Ketjoy, N.; Rakwichian, W.; Sukchai, S. Investigation on Temperature Coefficients of Three Types Photovoltaic Module Technologies under Thailand Operating Condition. *Procedia Eng.* **2012**, *32*, 376–383. [CrossRef]
5. Global EV Report. Available online: https://www.iea.org/reports/global-ev-outlook-2019 (accessed on 20 January 2021).
6. Chukwu, U.C.; Mahajan, S.M. V2G parking lot with PV rooftop for capacity enhancement of a distribution system. *IEEE Trans. Sustain. Energy* **2013**, *5*, 119–127. [CrossRef]
7. Honarmand, M.; Zakariazadeh, A.; Jadid, S. Self-scheduling of electric vehicles in an intelligent parking lot using stochastic optimization. *J. Frankl. Inst.* **2015**, *352*, 449–467. [CrossRef]
8. Heydarian-Forushani, E.; Golshan, M.E.; Shafie-khah, M. Flexible interaction of plug-in electric vehicle parking lots for efficient wind integration. *Appl. Energy* **2016**, *179*, 338–349. [CrossRef]
9. Song, Y.; Zheng, Y.; Hill, D.J. Optimal scheduling for EV charging stations in distribution networks: A Convexified model. *IEEE Trans. Power Syst.* **2016**, *32*, 1574–1575. [CrossRef]
10. Ivanova, A.; Fernandez, J.A.; Crawford, C.; Djilali, N. Coordinated charging of electric vehicles connected to a net-metered PV parking lot. In Proceedings of the 2017 IEEE PES Innovative Smart Grid Technologies Conference Europe (ISGT-Europe), Turin, Italy, 26–29 September 2017; pp. 1–6.
11. Figueiredo, R.; Nunes, P.; Brito, M.C. The feasibility of solar parking lots for electric vehicles. *Energy* **2017**, *140*, 1182–1197. [CrossRef]
12. Aghajani, S.; Kalantar, M. Operational scheduling of electric vehicles parking lot integrated with renewable generation based on bi-level programming approach. *Energy* **2017**, *139*, 422–432. [CrossRef]
13. Zhang, Y.; Cai, L. Dynamic charging scheduling for EV parking lots with photovoltaic power system. *IEEE Access* **2018**, *6*, 56995–57005. [CrossRef]
14. Yan, Q.; Zhang, B.; Kezunovic, M. Optimized operational cost reduction for an EV charging station integrated with battery energy storage and PV generation. *IEEE Trans. Smart Grid* **2018**, *10*, 2096–2106. [CrossRef]
15. Eldeeb, H.H.; Faddel, S.; Mohammed, O.A. Multi-objective optimization technique for the operation of grid tied PV powered EV charging station. *Electr. Power Syst. Res.* **2018**, *164*, 201–211. [CrossRef]
16. Jannati, J.; Nazarpour, D. Multi-objective scheduling of electric vehicles intelligent parking lot in the presence of hydrogen storage system under peak load management. *Energy* **2018**, *163*, 338–350. [CrossRef]
17. Chen, C.R.; Chen, Y.S.; Lin, T.C. Optimal Charging Scheduling for Electric Vehicle in Parking Lot with Renewable Energy System. In Proceedings of the 2019 IEEE International Conference on Systems, Man and Cybernetics (SMC), Bari, Italy, 6–9 October 2019; pp. 1684–1688.
18. Jiang, W.; Zhen, Y. A Real-time EV Charging Scheduling for Parking Lots with PV System and Energy Store System. *IEEE Access* **2019**, *7*, 86184–86193. [CrossRef]
19. Turan, M.T.; Ates, Y.; Erdinc, O.; Gokalp, E.; Catalão, J.P. Effect of electric vehicle parking lots equipped with roof mounted photovoltaic panels on the distribution network. *Int. J. Electr. Power Energy Syst.* **2019**, *109*, 283–289. [CrossRef]
20. Seddig, K.; Jochem, P.; Fichtner, W. Two-stage stochastic optimization for cost-minimal charging of electric vehicles at public charging stations with photovoltaics. *Appl. Energy* **2019**, *242*, 769–781. [CrossRef]

21.	Sedighizadeh, M.; Mohammadpour, A.; Alavi, S.M.M. A daytime optimal stochastic energy management for EV commercial parking lots by using approximate dynamic programming and hybrid big bang big crunch algorithm. *Sustain. Cities Soc.* **2019**, *45*, 486–498. [CrossRef]

22.	Ahmadi-Nezamabad, H.; Zand, M.; Alizadeh, A.; Vosoogh, M.; Nojavan, S. Multi-objective optimization based robust scheduling of electric vehicles aggregator. *Sustain. Cities Soc.* **2019**, *47*, 101494. [CrossRef]

23.	Wang, Z.; Jochem, P.; Fichtner, W. A scenario-based stochastic optimization model for charging scheduling of electric vehicles under uncertainties of vehicle availability and charging demand. *J. Clean. Prod.* **2020**, *254*, 119886. [CrossRef]

24.	Cao, Y.; Du, J.; Qian, X.; Nojavan, S.; Jermsittiparsert, K. Risk-involved stochastic performance of hydrogen storage based intelligent parking lots of electric vehicles using downside risk constraints method. *Int. J. Hydrog. Energy* **2020**, *45*, 2094–2104. [CrossRef]

25.	Cao, Y.; Huang, L.; Li, Y.; Jermsittiparsert, K.; Ahmadi-Nezamabad, H.; Nojavan, S. Optimal scheduling of electric vehicles aggregator under market price uncertainty using robust optimization technique. *Int. J. Electr. Power Energy Syst.* **2020**, *117*, 105628. [CrossRef]

26.	Bayrakcia, M.; Choib, Y.; Brownson, J. Temperature Dependent Power Modeling of Photovoltaics. *Energy Procedia* **2014**, *57*, 745–754. [CrossRef]

27.	Dash, P.K.; Gupta, N.C. Effect of Temperature on Power Output from Different Commercially available Photovoltaic Modules. *Int. J. Eng. Res. Appl.* **2015**, *5*, 148–151.

28.	Zhu, H.; Lian, W.; Lu, L.; Kamunyu, P.; Yu, C.; Dai, S.; Hu, Y. Online Modelling and Calculation for Operating Temperature of Silicon-Based PV Modules Based on BP-ANN. *Int. J. Photoenergy* **2017**, *2017*, 6759295. [CrossRef]

29.	Fesharaki, V.J.; Dehghani, M.; Fesharaki, J.J. The Effect of Temperature on Photovoltaic Cell Efficiency. In Proceedings of the 1st International Conference on Emerging Trends in Energy Conservation—ETEC, Tehran, Iran, 30–21 November 2011.

30.	Finnish Weather Data. Available online: https://www.ilmatieteenlaitos.fi/havaintojen-lataus (accessed on 20 January 2021).

31.	Sadati, S.M.B.; Moshtagh, J.; Shafie-Khah, M.; Rastgou, A.; Catalão, J.P. Optimal charge scheduling of electric vehicles in solar energy integrated power systems considering the uncertainties. In *Electric Vehicles in Energy Systems*; Springer: Cham, Switzerland, 2020; pp. 73–128.

32.	Prices for Electricity Products. Available online: https://www.vaasansahko.fi/en/prices-for-electricity-products-2/ (accessed on 15 December 2020).

33.	Nordpool Market Data. Available online: https://www.nordpoolgroup.com/ (accessed on 15 December 2020).

Article

Resilient Predictive Control Coupled with a Worst-Case Scenario Approach for a Distributed-Generation-Rich Power Distribution Grid

Nouha Dkhili [1,2], **Julien Eynard** [1,2], **Stéphane Thil** [1,2] and **Stéphane Grieu** [1,2,*]

1. PROMES-CNRS (UPR 8521), Rambla de la Thermodynamique, Tecnosud, 66100 Perpignan, France; nouha.dkhili@promes.cnrs.fr (N.D.); julien.eynard@univ-perp.fr (J.E.); stephane.thil@univ-perp.fr (S.T.)
2. Physical and Engineering Sciences Department, Université de Perpignan Via Domitia, 52 Avenue Paul Alduy, 66860 Perpignan, France
* Correspondence: grieu@univ-perp.fr; Tel.: +33-4-68-68-22-57

Abstract: In a context of accelerating deployment of distributed generation in power distribution grid, this work proposes an answer to an important and urgent need for better management tools in order to 'intelligently' operate these grids and maintain quality of service. To this aim, a model-based predictive control (MPC) strategy is proposed, allowing efficient re-routing of power flows using flexible assets, while respecting operational constraints as well as the voltage constraints prescribed by ENEDIS, the French distribution grid operator. The flexible assets used in the case study—a low-voltage power distribution grid in southern France—are a biogas plant and a water tower. Non-parametric machine-learning-based models, i.e., Gaussian process regression (GPR) models, are developed for intraday forecasting of global horizontal irradiance (GHI), grid load, and water demand, to better anticipate emerging constraints. The forecasts' quality decreases as the forecast horizon grows longer, but quickly stabilizes around a constant error value. Then, the impact of forecasting errors on the performance of the control strategy is evaluated, revealing a resilient behaviour where little degradation is observed in terms of performance and computation cost. To enhance the strategy's resilience and minimise voltage overflow, a worst-case scenario approach is proposed for the next time step and its contribution is examined. This is the main contribution of the paper. The purpose of the min–max problem added upstream of the main optimisation problem is to both anticipate and minimise the voltage overshooting resulting from forecasting errors. In this min–max problem, the feasible space defined by the confidence intervals of the forecasts is searched, in order to determine the worst-case scenario in terms of constraint violation, over the next time step. Then, such information is incorporated into the decision-making process of the main optimisation problem. Results show that these incidents are indeed reduced thanks to the min–max problem, both in terms of frequency of their occurrence and the total surface area of overshooting.

Keywords: smart grid paradigm; distributed generation; model-based predictive control; robustness; worst-case scenario; min–max optimisation; intraday forecasting; Gaussian process regression; machine learning

Citation: Dkhili, N.; Eynard, J.; Thil, S.; Grieu, S. Resilient Predictive Control Coupled with a Worst-Case Scenario Approach for a Distributed-Generation-Rich Power Distribution Grid. *Clean Technol.* **2021**, *3*, 629–655. https://doi.org/10.3390/cleantechnol3030038

Academic Editor: Lieven Vandevelde

Received: 28 May 2021
Accepted: 27 August 2021
Published: 30 August 2021

Publisher's Note: MDPI stays neutral with regard to jurisdictional claims in published maps and institutional affiliations.

1. Introduction

Worldwide, the transition to renewable-energy-based power generation is in full swing. Because power grids were originally designed for centralised power generation with unidirectional power flow, large-scale deployment of renewable energy technologies, hereinafter referred to as distributed generation, ushers in numerous operational issues. The concept of "smart grid" was born out of the need to better monitor the behaviour of these evolving power grids, to more accurately anticipate the operational issues that could be caused by new components, and to more efficiently control them to ensure safety and service quality.

Distributed power generators as well as storage devices are key players in modern power distribution grids. As a consequence, it goes without saying that these generators and devices can be used to balance power supply and demand. Based on this, the smart grid paradigm was mainly elaborated to tackle monitoring and control problems in power distribution grids [1]. Smart grids include advanced metering infrastructure and smart management schemes, which aim to boost grid observability and balance out power supply and demand, respectively. Regarding smart management schemes, optimal power flow (OPF) [2–7], demand-side management (DSM) [8–13], and multi-agent systems (MAS) [14–18] are some of the most numerous techniques used. One technique may be more appropriate than another, according to the target to be achieved, which can range from dimensioning and planning to real-time monitoring and control of power grids, and the technical and computational constraints to be taken into account. A survey of techniques for smart management of power distribution grids with prolific distributed generation is provided in [19]. The interested reader is referred to this paper.

This work falls within the scope of the "Smart Occitania" project (2017–2021), funded by the French agency for ecological transition (ADEME) and defined as a proof of concept for rural/suburban power distribution grids. An MPC-based strategy is proposed by PROMES-CNRS for smart management of a low-voltage power distribution grid with high levels of photovoltaics penetration using flexible assets, namely a biogas plant and a water tower. A simulated case study is carried out on a residential neighbourhood located in the Occitania region (southern France). The proposed strategy, which consists in an MPC scheme aiming to close the gap between power supply and demand, as stipulated by voltage constraints and the flexible assets' operational constraints [20], combines flexible asset management [21–23] and implicit model-based predictive control [24–27]. This strategy is fully explained in [20]; its principles are briefly outlined in Section 4.2. The suitability of model-based predictive control to the management of power distribution grids subject to disturbances—intermittent renewable-energy-based power generation and stochastic power demand—is plain. Because of their weakly meshed (often radial) structure, low-voltage power distribution grids are especially prone to cascading failures provoked by disturbances.

The main contribution of PROMES-CNRS is the a new problem formulation allowing the mixed-integer nonlinear programming (MINLP) [28–34] setting due to the ON/OFF controller the water tower is equipped with—this setting greatly increases the computational complexity of the optimisation problem—to be solved as a smooth continuous one without resorting to relaxation [20]. An in-depth analysis of the results has highlighted the MPC strategy's potential for upper-level power flow management and curtailment of voltage fluctuations in the considered low-voltage power distribution grid. Clearly, the strategy is a step towards low-voltage power distribution grids capable of integrating renewable-energy-based power generation while maintaining stability and quality of service.

The proposed MPC scheme incorporates intraday forecasts of grid load, water demand, and global horizontal irradiance (from which PV power generation is inferred; global horizontal irradiance is the total amount of shortwave radiation received from above by a surface horizontal to the ground), as well as their associated confidence intervals, in order to efficiently operate the aforementioned flexible assets towards balancing supply and demand within the grid. It does so using Gaussian process regression models. The stochasticity of the controller's inputs poses a threat to its efficacy, since it can incur unforeseen constraint violation (here, in the form of voltage undershooting/overshooting). The issue of stochastic data in MPC schemes is present in a wide array of problems, which makes it an active research field. There exists a significant body of literature handling stochastic MPC for linear problems [35–37].

As for nonlinear problems, which is the case addressed in this paper, more recent works have tackled them in different ways. A common rationale in the scientific literature is "multi-stage" control schemes. Recent examples include offline computation of an incremental Lyapunov function, which is then used for an online construction of a "tube" to tighten the constraints [38], a decomposition into several deterministic sub-problems whose solutions are then aggregated using an operation-cost-based rule [39], and the modelling of the uncertainty through a tree of discrete scenarios, coupled with a tube-based MPC to balance the system's variability and its economic profitability [40,41].

The use of a multi-stage approach adds a layer of complexity into the control scheme. The advantages of such methods are their ability to combine several different techniques into a hierarchical scheme to tackle numerous difficulties in the problem. This often presents itself as an offline stage that feeds into an online one. Their obvious drawbacks is the added complexity and, in the case of scenario-based methods, significant computational burden, which make them ill-suited for real-time applications.

The method proposed in this paper is based on min–max MPC for uncertain nonlinear systems under constraints [42,43]. This technique's premise boils down to risk aversion by computation of a worst-case scenario upstream of a standard optimisation problem. The main optimisation problem upon which the predictive control strategy is based seeks optimal flexible assets' setpoints, in order to reduce the gap between power supply and demand in the power distribution grid. The solution proposed here adds a layer, to be called "the min-max problem", upstream of this main optimisation problem. The min–max problem will determine the values of possible PV power generation, grid load, and water demand for which constraint violation will be at a maximum, within the forecasts' respective confidence intervals. The scenario with these values is dubbed the worst-case scenario. Then, the predictive controller searches for optimal flexible assets' setpoints that would uphold constraints computed with values of PV power generation, grid load, and water demand corresponding to the worst-case scenario. The controller searches for worst-case scenario within a feasible space defined by the aforementioned confidence intervals associated to the GPR forecasts.

The paper is organised as follows: Section 2 presents the case study treated in this work. Section 3 provides a definition of Gaussian process regression, kernel compositions of the models developed to forecast all three stochastic quantities, and the obtained results. Then, Section 4 details the proposed model-based predictive control strategy, starting with a step-by-step explanation of the strategy's inner-workings, formulating the main optimisation problem upon which the strategy is based, and introducing the worst-case scenario approach which enhances the robustness of the control strategy to forecasting errors. In Section 5, the impact of forecasting errors on the performance of the model-based predictive control strategy is analysed and an evaluation is carried out of the contribution of the worst-case approach to enhancing the scheme's robustness to forecasting errors. Section 6 recapitulates the main findings of the paper and discusses possible improvements.

2. Case Study

2.1. Low-Voltage Power Distribution Grid

The case study [20] is carried out on a simulated low-voltage power distribution grid (a residential neighbourhood) in the Occitania region in southern France, composed of approximately 600 households, 200 household PV installations of 4 kW each, amounting to a total capacity of 800 kW, a small-scale biogas plant (100 kW), and a water tower (100 kW). Figure 1 displays some grid load and PV power generation data (over a week in April) used in this work. Grid load is measured at the MV/LV transformer level of the low-voltage power distribution grid whereas PV power generation is inferred from GHI measurements taken by a pyranometer installed at PROMES-CNRS laboratory, which is located a few kilometres from the residential neighbourhood.

In this section, models of the power distribution grid and the flexible assets used in this case study are presented, formulated over a forecast horizon H. Let H_p be the integer number of time slots within this forecast horizon. In the following, and for all time-dependant quantities, $t \in \{1, \ldots, H_p\}$.

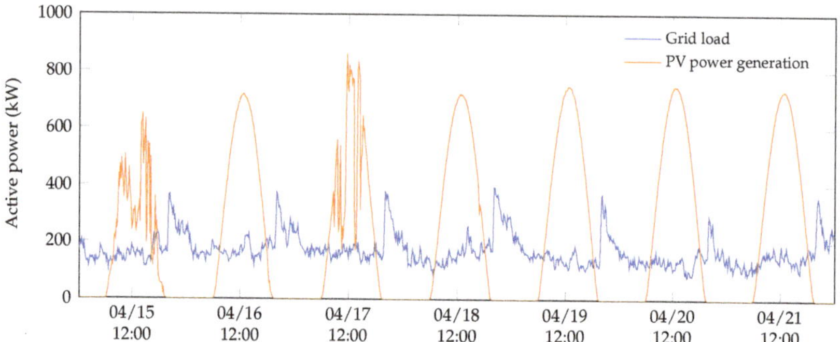

Figure 1. Grid load and PV power generation over a week in April (case study).

Measurements made at the MV/LV transformer level of the power distribution grid and used throughout this study correspond to means over each time slot (herein, a 10-min time step is considered). This study focuses on upholding contractual voltage bounds; voltage means must at all times remain within $\pm\ \delta U$ of the nominal value, i.e., $\forall q \in \{1,\ldots,N\}$ and $\forall t \in \{1,\ldots,H_p\}$:

$$|U_q(t) - U_n| \leqslant \delta U \tag{1}$$

where U_n is the nominal single-phase voltage value for all grid nodes, N is the number of nodes in the low-voltage power distribution grid, H_p is the integer number of time slots within the forecast horizon, and δU is the acceptable margin of voltage variations with respect to the nominal value. The acceptable margin fixed by the French distribution grid operator ENEDIS is 10% of the nominal value. However, for the purposes of this study, a lower margin of 3% is considered in order to better flesh out possible voltage overflow phenomena due to power supply/demand unbalance within the grid.

2.2. Flexible Assets

In this section, the biogas plant and water tower models are briefly described. Details about these models and the characteristics of the flexible assets are given in [20].

2.2.1. Biogas Plant

The biogas volume in the storage unit (in m^3) is described as:

$$V_b(t+1) = V_b(t) + \frac{T}{60}\left(Q_{b,in}(t) - \frac{P_b(t)}{\eta \cdot \text{LHV}}\right) \tag{2}$$

where T is the time step ($T = 10\,\text{min}$), $Q_{b,in}$ is the flow rate of biogas production entering the storage unit (in $\text{m}^3\,\text{h}^{-1}$), P_b is the plant's active power output (in W), η is the generator's efficiency, and LHV is the lower heating value of the stored biogas (in $\text{kWh}\,\text{m}^{-3}$). The output P_b is subject to the following constraint:

$$P_{b,min} \leqslant P_b(t) \leqslant P_{b,max} \tag{3}$$

where $P_{b,min}$ and $P_{b,max}$ are the minimum and maximum power generation of the biogas plant, respectively.

The biogas volume in the storage unit is subject to the following constraint:

$$V_{b,min} \leqslant V_b(t) \leqslant V_{b,max} \tag{4}$$

where $V_{b,min}$ and $V_{b,max}$ are the minimum and maximum biogas storage capacities of the biogas plant, respectively.

2.2.2. Water Tower

The water volume in the storage tank (in m^3) is described as follows:

$$V_w(t+1) = V_w(t) + \frac{T}{60}\left(\frac{3600\eta_w}{\rho \cdot g \cdot h}P_w(t) - Q_{w,out}(t)\right) \tag{5}$$

where T is the time step ($T = 10\,\text{min}$), $Q_{w,out}$ is the flow rate of water demand (in m^3 h^{-1}), P_w is the water pump's active power consumption (in W), η_w is the water pump's efficiency, ρ is the water density (in kg m^{-3}), g is the gravitational acceleration (in m s^{-2}), and h is the water level in the storage tank (in m).

Let $P_{w,min}$ and $P_{w,max}$ be the minimum and maximum power consumption values of the water tower, respectively. The power consumption P_w can only be set following ON/OFF commands, i.e., it is subject to the following constraint:

$$P_w(t) \in \{P_{w,min}; P_{w,max}\} \tag{6}$$

The water volume in the storage tank is subject to the following constraint:

$$V_{w,min} \leqslant V_w(t) \leqslant V_{w,max} \tag{7}$$

where $V_{w,min}$ and $V_{w,max}$ are the minimum and maximum storage capacities of the water tank, respectively.

2.3. PV Power Generation Inferred from Global Horizontal Irradiance

PV power generation values are inferred from global horizontal irradiance values using the following Equation [44]:

$$\widehat{P}_{PV}(t) = \eta_{T_{ref}} \cdot S \cdot \widehat{GHI}(t) \cdot \tau_\alpha\left[1 - \beta_{ref}\left(T_p - T_{ref}\right)\right] \tag{8}$$

where $\eta_{T_{ref}}$ (herein, $\eta_{T_{ref}} = 0.21$) is the PV panel's efficiency, S is the total surface area of PV panels in the power distribution grid, $\widehat{GHI}$ are global horizontal irradiance forecasts, τ_α is the effective transmittance of the PV panels (herein, $\tau_\alpha = 0.95$), β_{ref} is the coefficient of power degradation due to high temperatures ($\beta_{ref} = 0.004$), T_{ref} is the reference temperature (herein, $T_{ref} = 25\,°C$), and T_p is the PV panels' temperature, computed as follows [45]:

$$T_p(t) = T_a(t) + k \cdot \widehat{GHI}(t) \tag{9}$$

where T_a is the ambient temperature and $k = 0.025$.

2.4. Initial Operation Strategy

The initial operation of the flexible assets does not take into account power grid regulation purposes [20]. The biogas plant's power generation is a constant nominal value (herein, $P_{b,n} = 100\,\text{kW}$). This is in line with the steady biogas flow generated by the bioreactor and coming into the storage unit. As for the water tower, its water pump is subject to an ON/OFF controller, which ensures that the water level remains between two threshold values at all times (herein, $P_{w,n} = 100\,\text{kW}$). Both assets' priority is maintaining storage volume levels within pre-fixed extrema, which can sometimes be in conflict with the grid stability's best interest.

3. Intraday Forecasting of Stochastic Quantities

3.1. Introduction

In the context of the control strategy proposed in this paper, the three stochastic quantities that come into play are the following: power grid load, global horizontal irradiance (GHI) [46–48], and water demand. The power grid load represents agglomerated power demand of households in the studied suburban area. It is the grid operator's priority to make sure this demand is met at all times, under adequate quality and security standards. PV power generation, inferred from GHI values as explained in Section 2.3, represents the agglomerated power generation of household PV panels in the studied area (the residential neighbourhood). One of the assets used by the control strategy is the water tower, whose operational priority is meeting the water demand. In this paper, the methodology used to forecast each of these quantities and its results are briefly presented.

As part of the research activities of the PROMES-CNRS laboratory, a comparative study of models developed for multi-step-ahead GHI forecasting is carried out for intrahour and intraday forecast horizons using a two-year database of GHI measurements sampled at a 10-min rate. Results demonstrate that, although all considered models outperform the persistence model, there is no clear frontrunner in terms of nRMSE values, with a slight advantage for LSTM (long short term memory) artificial neural network (ANN) and Gaussian process regression (GPR) models when taking into account the quality of the forecasts' associated confidence intervals.

Confidence intervals are a noteworthy perk of using GPR models in the forecast module of the smart management strategy since they are built-in in the models and do not require running a Monte Carlo simulation to be statistically inferred as is the case for other forecasting methods (like artificial neural networks). These confidence intervals give the predictive controller supplementary information it uses to achieve a more robust control strategy. The development of GPR models for intraday GHI forecasting are detailed in [46–48].

GPR models used herein are, in part due to the fact that the database used to update the model's parameters is a sliding one, sufficiently fast to be a valid candidate for the real-time control application at hand. On the downside, the predictive nature of the controller makes it dependant upon real-time access to measurements to be able to sustain the forecasting models' sliding databases and to update their parameters at each time step. This limitation is true for all machine-learning-based methods operating in real time, hence the growing interest in the development of state-of-the-art smart metering technologies that ensure quick and reliable data transfer.

For the reasons listed above, the choice has been made to develop GPR models to provide intraday forecasts for the smart management scheme developed herein. The associated confidence intervals are incorporated into the control scheme to improve its robustness to forecasting errors as explained in Section 4. This is the main contribution of the paper as it focuses on the developement of a resilient predictive control strategy for low-voltage power distribution grids with prolific distributed generation.

3.2. Gaussian Process Regression

A Gaussian process (GP) can be seen as a collection of random variables, any finite number of which have a joint Gaussian distribution [48,49]. A prior over functions is defined and can then be converted into a posterior over functions once some data has been observed (i.e., observations). $f(x) \sim \mathcal{GP}(\mu(x), k(x, x'))$, with x and x' arbitrary input variables, $\mu(x) = \mathbb{E}[f(x)]$ the mean function (which is usually assumed to be null) and $k(x, x') = \mathbb{E}[(f(x) - \mu(x))(f(x') - \mu(x')^\mathsf{T})]$ the covariance function (also known as *kernel*), indicates that this random function follows a GP. The interested reader is referred to [49] for a detailed explanation of Gaussian processes and Gaussian process regression (GPR).

The standard regression model with additive noise is formulated as follows:

$$y = f(x) + \varepsilon \tag{10}$$

where $x \in \mathbb{R}^{D \times 1}$ is the input vector with a dimension of 1, f is the regression function, y is the observed value and $\varepsilon \sim \mathcal{N}(0, \sigma_\varepsilon^2)$ is an independent, identically distributed Gaussian noise.

The mean prediction μ_* can be written as a linear combination of kernel functions, each one centred on a data point:

$$\mu_* = \mu(x_*) + k_*^{\mathsf{T}} \cdot \alpha = \mu(x_*) + \sum_{i=1}^{n} \alpha_i \cdot k(x_i, x_*) \tag{11}$$

where $\alpha = \left(K + \sigma_\varepsilon^2 \cdot I\right)^{-1}(y - \mu(x_*))$. The coefficients α_i are referred to as parameters.

It is possible to combine several kernel functions to obtain a more complex one. The only requirement is that the resulting covariance matrix must be a positive semi-definite function. A detailed list of kernel functions is provided in [49]. Hereinafter, the kernel functions used in this work are defined.

The periodic kernel (Per) is given by:

$$k_{\mathrm{Per}}(x, x') = \sigma_1^2 \cdot \exp\left(-\frac{2\sin^2\left(\frac{\pi(x-x')}{P}\right)}{\ell_1^2}\right) \tag{12}$$

where σ_1 is the amplitude, ℓ_1 is the correlation length and P is the period.

The isotropic squared exponential (SE) kernel is given by:

$$k_{\mathrm{SE}}(x, x') = \sigma_2^2 \cdot \exp\left(-\frac{(x-x')^2}{2\ell_2^2}\right) \tag{13}$$

where σ_2 is the amplitude and ℓ_2 is the correlation length.

The isotropic rational quadratic (RQ) kernel is given by:

$$k_{\mathrm{RQ}}(x, x') = \sigma_3^2 \left(1 + \frac{(x-x')^2}{2\ell_3^2 \cdot \alpha}\right)^{-\alpha} \tag{14}$$

where σ_3 is the amplitude, ℓ_3 is the correlation length and α defines the relative weighting of large-scale and small-scale variations.

In addition to forecasts, the regression model provides an associated confidence interval within which measurements have a probability of 95% of staying. This interval ($\mathbb{CI}$) is computed as follows:

$$\mathbb{CI} = \mu_* \pm 1.96\sqrt{\sigma_*^2} \tag{15}$$

where $\mathbb{CI}$ represent the confidence interval bounds, μ_* is the predictive mean, and σ_* is the predictive variance.

During the training phase, the values of hyperparameters are optimised using a 2-week database [48]. During the test phase, at every time step, new measurements are incorporated in order to update values of the model's parameters using a sliding 24 h database.

3.3. Data Description and Kernel Compositions

The forecast horizons considered in this work are intraday: they range from 1 h to 24 h. For the training phase, two weeks of data are used. The testing phase is then performed over one week. Available data of power grid load, water demand, and global horizontal irradiance span a year for the three stochastic quantities. They are sampled at a 10-min rate. An examination of the signals' behaviours informs the following choices of kernel compositions for their intraday forecasting.

1. Power grid load

$$k^{gl}(x, x') = k_{\mathrm{Per}}^{gl}(x, x') + k_{\mathrm{SE}}^{gl}(x, x') + k_{\mathrm{RQ}}^{gl}(x, x') \tag{16}$$

Herein, the period of the data is 24 h and the structure of k_{Per}^{gl} translates the daily periodic pattern. k_{SE}^{gl} is used to capture the long-term trend of the data, namely the seasonal tendencies present in the power grid load. k_{RQ}^{gl} is used to capture the intraday variations observed in the data, namely intraday fluctuations due to behavioural patterns of end-users.

2. Water demand

$$k^{wt}(x, x') = k_{\text{Per}}^{wt}(x, x') + k_{\text{SE}}^{wt}(x, x') \tag{17}$$

k_{Per}^{wt} models the daily periodic shape of the data and k_{SE}^{wt} is used to fit the data's intraday fluctuations. A brief kernel study leads to the conclusion that the simple addition of a squared exponential kernel is sufficient to have satisfactory results without adding too much computational burden to the model.

3. Global horizontal irradiance

$$k^{\text{GHI}}(x, x') = k_{\text{Per}}^{\text{GHI}}(x, x') \cdot k_{\text{RQ}}^{\text{GHI}}(x, x') \tag{18}$$

$k_{\text{Per}}^{\text{GHI}}$ models the daily periodic shape of the data and $k_{\text{RQ}}^{\text{GHI}}$ captures the intraday fluctuations present in these data. A thorough study has been conducted in [47] to determine the most suitable kernel composition for intraday GHI forecasting.

3.4. Forecasting Results

The evaluation metrics used in this work, evaluated over the considered week in April (see Figure 1), are the following:

1. The normalized root mean square error (nRMSE), expressed as follows:

$$\text{nRMSE} = 100 \frac{\sqrt{\frac{1}{n_*} \sum_{i=1}^{n_*} \left(y_{\text{test}}(i) - y_{\text{forecast}}(i)\right)^2}}{\frac{1}{n_*} \sum_{i=1}^{n_*} y_{\text{test}}(i)} \tag{19}$$

where $y_{\text{test}} \in \mathbb{R}^{n_* \times 1}$ are the test data, $y_{\text{forecast}} \in \mathbb{R}^{n_* \times 1}$ are the forecasts given by the models, and n_* is the number of data points in the forecast horizon.

2. The coverage width-based criterion (CWC) [50], defined as a combination of two different criteria, i.e., the prediction interval normalized average width (PINAW) and the prediction interval coverage probability (PICP).
The PINAW criterion allows the surface area of the confidence intervals associated with the forecasts to be assessed:

$$\text{PINAW} = \frac{1}{n \cdot R} \sum_{i=1}^{n} (U_i - L_i) \tag{20}$$

where R is the difference between the maximum and minimum in test data and U_i and L_i are the upper and lower bounds of the confidence interval, respectively.
The PICP criterion informs on the probability that measurements would fall within the confidence interval:

$$\text{PICP} = \frac{1}{n} \sum_{i=1}^{n} \epsilon_i \tag{21}$$

where ϵ_i is used to detect whether the target value is in the confidence interval.
The coverage width-based criterion (CWC) is then defined as follows:

$$\text{CWC} = \text{PINAW}\left(1 + \gamma(\text{PICP}) \cdot \exp(-\eta \cdot (\text{PICP} - \mu))\right) \tag{22}$$

where η $(\eta = 10)$ and μ $(\mu = 0.95)$ are parameters, and γ is defined such that:

$$\gamma(x) = \begin{cases} 1 & \text{if } x < \mu \\ 0 & \text{if } x \geqslant \mu \end{cases} \tag{23}$$

Figure 2 displays nRMSE values of intraday forecasts of power grid load, water demand, and GHI. For intraday forecasts of grid load, results show that the GPR model performs well for short time horizons. However, the forecasting error rapidly increases as the forecast horizon does, which is an expected result. For a 1-h horizon, the GPR model achieves an error as low as 13%. As for the 6-h horizon and the 12-h one, nRMSE values are virtually constant (21% and 23%, respectively). This remains the case for a forecast horizon of 24 h, where the GPR model has an error of 26%.

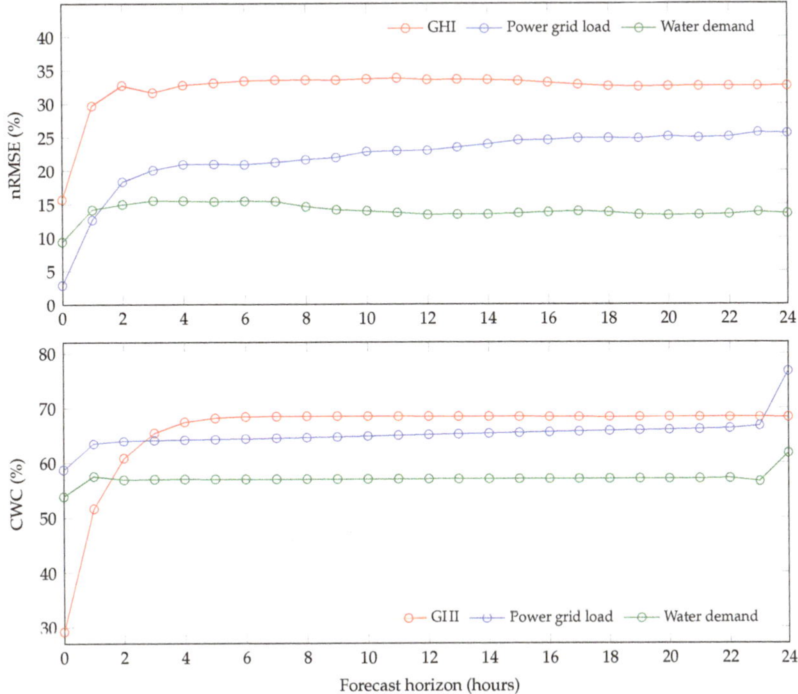

Figure 2. Evaluation criteria (nRMSE (19) and CWC (22)) for intraday GPR forecasts of GHI, power grid load and water demand, with respect to the forecast horizon, over a one-week period.

For intraday forecasts of water demand, values of nRMSE start at 9% for a 10 min forecast and quickly stabilises around 13% for longer forecast horizons. Due to the data's regular nature, a simple combination of a periodic kernel and a rational quadratic kernel for intraday fluctuations proves enough to capture the data's behaviour quite faithfully. As for intraday forecasts of GHI (from which PV power generation is inferred), the selected GPR model performs well for short forecast horizons (15.74% for a 10-min forecast horizon), but its performance degrades as the forecast horizon grows and it stabilises around 32% for horizons beyond 6 h. This is reflected in the temporal evolution of the forecasts, where it can be seen that intraday fluctuations become increasingly difficult to fit. Further work is currently underway to enhance the model's performance for long intraday forecast horizons.

Figure 2 also displays values of CWC, which follow similar patterns to those of nRMSE values: starting with lower values for short forecast horizons, then quickly stabilizing around a constant value for longer horizons. It can be seen that the higher the nRMSE values, the higher the corresponding CWC values. They stabilize around 57% for water demand forecasts, around 65% for grid load forecasts, and around 68% for GHI forecasts.

4. Model-Based Predictive Control Strategy

4.1. Inner-Workings of the Smart Management Scheme

Figure 3 shows the synoptic scheme of the proposed MPC strategy. The following is a detailed explanation of the different steps taken by the proposed strategy at each time step.

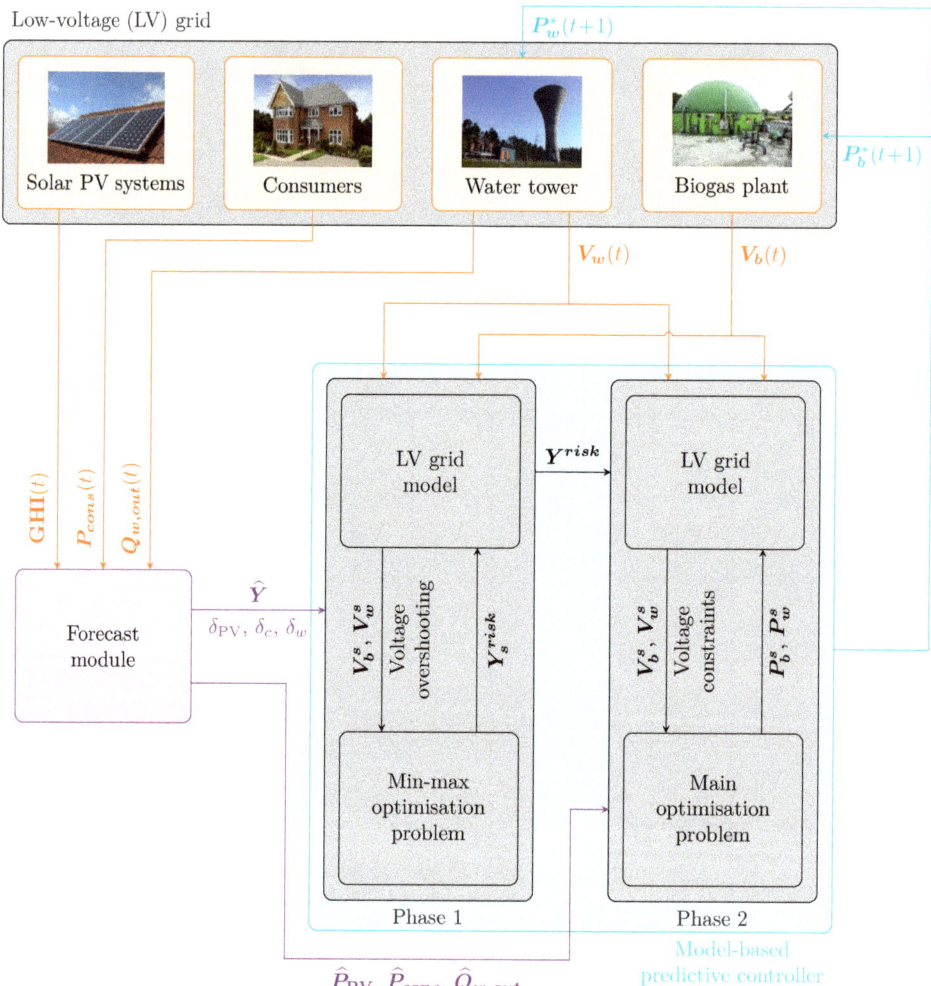

Figure 3. Synoptic scheme of the amended MPC-based strategy for smart management of a low-voltage power distribution grid using flexible assets. Let **GHI**, P_{cons}, $Q_{w,out}$, V_w, V_b be measurements of global horizontal irradiance, grid load, water demand, water volume, and biogas volume, respectively. Let $\widehat{Y}$ be forecasts of stochastic inputs for the following time steps. Let δ_{PV}, δ_c, δ_w be margins that define confidence intervals for the next time step of GPR forecasts of PV power generation, grid load, and water demand, respectively. Let $\widehat{P}_{PV}$, $\widehat{P}_{cons}$, $\widehat{Q}_{w,out}$ be forecasts of PV power generation, grid load, and water demand over the forecast horizon, respectively. Let Y_s^{risk} and Y^{risk} be candidate values and optimal values of worst-case scenario stochastic inputs. Let P_b^s and P_w^s be candidate values of biogas plant setpoints and water tower setpoints, respectively. Let V_b^s and V_w^s be the biogas volume and the water volume, respectively, corresponding to the candidate optimisation variables of a given iteration.

1. Data acquisition: measurements are taken of stored biogas volume and stored water volume and are injected into the predictive controller in order to update the system's

state. GHI and grid load are measured and water demand is inferred from measurements of water volume and incoming flow rate $Q_{w,in}$. These values are injected into the forecast module.

2. Forecast module: measured values of the controller's stochastic inputs (GHI, grid load, and water demand inferred from incoming flow rate) are added to the sliding databases of respective GPR models, which are then used to update the models' parameters. Afterwards, the module produces updated forecasts of all three stochastic quantities over the forecast horizon, along with their respective confidence intervals. Lastly, GHI values and their corresponding confidence intervals are converted into PV power generation ones.

3. Worst-case scenario: phase 1 of the controller's decision-making process consists in determining the stochastic input values corresponding to the worst-case scenario of the following time slot, in terms of constraint violation. This bloc receives measurements of biogas volume and water volume, GPR forecasts of PV power generation, grid load, and water demand for the following time slot, and confidence intervals of GPR forecasts for the following time slot.

   An optimisation algorithm searches the feasible space defined in Figure 4 for worst-case scenario input values. It does so based on the optimisation problem formulated by Equations (46)–(49) and using the low-voltage grid model to evaluate volume bounds and potential voltage overshooting.

4. Reduction of power supply/demand unbalance: phase 2 of the controller's decision-making process consists in determining the flexible assets' optimal setpoints. This bloc receives GPR forecasts of PV power generation, grid load, and water demand over the entire forecast horizon, as well as worst-case scenario input values of the following time slot, produced by phase 1.

   An optimisation algorithm searches for optimal setpoints of biogas plant power generation and water tower power consumption. It does so based on the optimisation problem defined in Section 4.2 and using the low-voltage grid model based on Kirchhoff laws to evaluate various constraints.

5. Implementation of flexible assets' setpoints: optimal setpoints of flexible assets are produced, the first step of which are implemented by the biogas plant and the water tower.

4.2. Main Optimisation Problem

The optimisation problem [20] is formulated such as the discrete setpoint values of the water tower are replaced with a real-values optimisation variable $\bar{t} \in \mathbb{R}^{H_p}$, with H_p the integer number of time slots within the forecast horizon, effectively avoiding the MINLP setting [28–34] without relaxing the problem. In fact, between sampling times t_i and t_{i+1}, the water tower can switch between its two states of operation ($P_{w,min}$ and $P_{w,max}$) and the biogas plant can switch between two setpoints within the interval $[P_{b,min}, P_{b,max}]$. If the assets do make the switch within a given time slot, they do so at the same instant $\bar{t}_i$. Let us denote X as follows:

$$X = \left[P_{b,\text{ON}} \; P_{b,\text{OFF}} \; \bar{t} \; U_{\text{ON},q} \; U_{\text{OFF},q} \right]^\mathsf{T} \tag{24}$$

where $P_{b,\text{ON}} \in \mathbb{R}^{H_p}$ and $P_{b,\text{OFF}} \in \mathbb{R}^{H_p}$ form the biogas plant setpoint like this:

$$P_b(\tau) = \begin{cases} P_{b,\text{ON}}(\tau), \tau \in [t_i, t_i + \bar{t}_i] \\ P_{b,\text{OFF}}(\tau), \tau \in [t_i + \bar{t}_i, t_{i+1}] \end{cases} \tag{25}$$

For every node $q \in \{1, \ldots, N\}$, $U_{\text{ON},q} \in \mathbb{R}^{(H_p \cdot N)}$ and $U_{\text{OFF},q} \in \mathbb{R}^{(H_p \cdot N)}$ form the voltages in the grid:

$$U_q(\tau) = \begin{cases} U_{\text{ON},q}(\tau), \tau \in [t_i, t_i + \bar{t}_i] \\ U_{\text{OFF},q}(\tau), \tau \in [t_i + \bar{t}_i, t_{i+1}] \end{cases} \tag{26}$$

The optimisation problem can be solved at each time step assuming that the first state of the water tower is always ON. In some extreme cases, this assumption may induce some issues of implementability given volume constraints, which are tackled in a post-treatment phase (the interested reader is referred to [20] for details about the post-treatment algorithm). However, this simplification reduces the complexity of the model without sacrificing much performance. The objective function f_{obj} is formulated as follows:

$$f_{obj}(\boldsymbol{X}) = \sum_{i=0}^{H_p-1} \left[\int_{t_i}^{t_i+\bar{t}_i} S_{\textbf{ON}}(\tau)\mathrm{d}t + \int_{t_i+\bar{t}_i}^{t_{i+1}} S_{\textbf{OFF}}(\tau)\mathrm{d}t \right] \tag{27}$$

with

$$S_{\textbf{ON}}(\tau) = |\boldsymbol{P}_{\textbf{PV}}(\tau) + \boldsymbol{P_b}(\tau) - P_{cons}(\tau) - P_{w,max}|^2 \tag{28}$$

$$S_{\textbf{OFF}}(\tau) = |\boldsymbol{P}_{\textbf{PV}}(\tau) + \boldsymbol{P_b}(\tau) - P_{cons}(\tau) - P_{w,min}|^2 \tag{29}$$

The optimisation problem is then formulated as follows:

$$\boldsymbol{X}^* = \arg\min_{\boldsymbol{X}} f_{obj}(\boldsymbol{X}) \tag{30}$$

and is subject to the following bounds and constraints, $\forall i \in \{1, \ldots, H_p\}$:

- Biogas plant power bounds

$$P_{b,min} \leqslant \boldsymbol{P_{b,\textbf{ON}}}(\tau) \leqslant P_{b,max} \tag{31}$$

$$P_{b,min} \leqslant \boldsymbol{P_{b,\textbf{OFF}}}(\tau) \leqslant P_{b,max} \tag{32}$$

- Switch time bounds

$$0 \leqslant \bar{t}_i \leqslant T \tag{33}$$

- Biogas volume constraints

$$V_{b,min} \leqslant \boldsymbol{V_b}(t) \leqslant V_{b,max} \tag{34}$$

- Water volume constraints

$$V_{w,min} \leqslant \boldsymbol{V_w}(t) \leqslant V_{w,max} \tag{35}$$

- Voltage constraints

$$\bar{t}_i \cdot K_t(\boldsymbol{P_{b,\textbf{ON}}}(\tau), P_{w,max}, \boldsymbol{U_{\textbf{ON},q}}(\tau)) = 0 \tag{36}$$

$$(T - \bar{t}_i) \cdot K_t(\boldsymbol{P_{b,\textbf{OFF}}}(\tau), P_{w,min}, \boldsymbol{U_{\textbf{OFF},q}}(\tau)) = 0 \tag{37}$$

$$|\boldsymbol{U_{\textbf{ON},q}}(\tau) - U_n| \leqslant \delta U \tag{38}$$

$$|\boldsymbol{U_{\textbf{OFF},q}}(\tau) - U_n| \leqslant \delta U \tag{39}$$

K_t describes voltage variations across the low-voltage power distribution grid as a function of the power which is injected or absorbed at each node. Kirchhoff's law constraints are presented as two sets of constraints (Equations (36) and (37)), which guarantee that Kirchhoff's laws are upheld in both sub-intervals of each time slot. The equation set depicting voltage variations is multiplied by $\bar{t}_i$ (Equation (36)) or $T - \bar{t}_i$ (37), using appropriate values of biogas plant and water tower setpoints for each interval, in order to ensure that only one constraint is activated in case of extreme values of $\bar{t}_i$. In fact, in case $\bar{t}_i = 0$, Equation (36) is ignored, reflecting the fact that during time slot i, the water tower is turned off immediately at the beginning of the time slot. Likewise, in case $\bar{t}_i = T$, Equation (37) is ignored since the water tower remains on for the duration of the time slot. Voltage constraints are also written as two sets of constraints (Equations (38) and (39)) that account for voltage variations in both states of the low-voltage power distribution grid within each time slot. While depicting the same physical constraints, this problem formulation has a bigger feasible set than the MINLP one [28–34]. As a result, with such a formulation, the global optimum is guaranteed to be equal or better than the one of the MINLP formulation.

4.3. The Worst-Case Scenario Approach

In this section, amendments are made to the predictive control strategy in order to enhance the controller's performance by making it more robust to forecasting errors of the system's various stochastic quantities. The premise of the method is to use min–max optimisation in order to find and solve a "worst-case scenario" at each time step based on confidence intervals given by the inputs' respective GPR models. Eliminating, or at least minimising, the constraint violations corresponding to the worst-case scenario guarantee that these violations are also minimised for all other possible scenarios.

It should be noted that, at each time step, these amendments are only made to the decision-making of the subsequent time step, and not over the entire forecast horizon. This choice is motivated by two reasons. The first is that determining a worst-case scenario over the entire forecast horizon is a conservative and computationally expensive optimisation problem, which is incompatible with the real-time application at hand. As a matter of fact, a min–max problem over the entire forecast horizon has a feasible space of dimension $(3 \times H_p)$, as opposed to the problem concerned only with the following time step only having a 3-dimensional feasible space. The second is that, due to the closed-loop nature of MPC, computing robust setpoints for the entire forecast horizon is in high likelihood a waste of resources because, at each time step, only the first setpoint is applied and the whole procedure is reiterated at the next one. Ergo, the scheme only seeks to provide a setpoint robust to forecasting errors for the next time step.

First, let $\hat{P}_{PV}$, $\hat{P}_{cons}$, and $\hat{Q}_{w,out}$ be forecasted values of PV power generation, grid load, and water demand, respectively, over the forecast horizon. Then, let P_{PV}, P_{cons}, and $Q_{w,out}$ be measured values of PV power generation, grid load, and water demand, respectively, over the forecast horizon. Finally, δ_{PV}, δ_c, and δ_w define confidence intervals, for the next time step, of GPR forecasts of PV power generation, grid load, and water demand, respectively, as follows:

$$P_{PV}(t+1) \in [\hat{P}_{PV}(t+1) - \delta_{PV}, \hat{P}_{PV}(t+1) + \delta_{PV}] \tag{40}$$

$$P_{cons}(t+1) \in [\hat{P}_{cons}(t+1) - \delta_c, \hat{P}_{cons}(t+1) + \delta_c] \tag{41}$$

$$Q_{w,out}(t+1) \in [\hat{Q}_{w,out}(t+1) - \delta_w, \hat{Q}_{w,out}(t+1) + \delta_w] \tag{42}$$

Herein, the confidence intervals are computed such that there is a 95% probability of measurements remaining inside them (15). There exists a triplet $(P_{PV}^{risk}(t+1), P_{cons}^{risk}(t+1), Q_{w,out}^{risk}(t+1))$, contained in the feasible space illustrated in Figure 4, that induces a worst-case scenario vis-a-vis the optimisation problem constraints in the next time step. Finding this triplet and incorporating it into the predictive control strategy described in Section 4 allows the controller to adjust its decision-making in order to reduce, if possible eliminate, any constraint violation that could occur in the next time step as a result of the stochastic quantities' measured values being different from the forecasted ones, within the limits of the associated confidence intervals. Let Y be the vector containing measurements of PV power generation, grid load, and water demand, for the next time step:

$$Y = \left[P_{PV}(t+1) \; P_{cons}(t+1) \; Q_{w,out}(t+1) \right]^T \tag{43}$$

Let $\widehat{Y}$ be the vector containing forecasts of PV power generation, grid load, and water demand, for the next time step:

$$\widehat{Y} = \left[\widehat{P}_{\text{PV}}(t+1)\ \widehat{P}_{cons}(t+1)\ \widehat{Q}_{w,out}(t+1)\right]^T \tag{44}$$

Let Y^{risk} be the vector comprised of critical values defining the worst-case scenario of the next time step:

$$Y^{risk} = \left[P_{\text{PV}}^{risk}(t+1)\ P_{cons}^{risk}(t+1)\ Q_{w,out}^{risk}(t+1)\right]^T \tag{45}$$

where P_{PV}^{risk}, P_{cons}^{risk}, and $Q_{w,out}^{risk}$ are critical values of PV power generation, grid load, and water demand, respectively, corresponding to the worst-case scenario.

Based on Equations (40)–(42), Y^{risk} exists in the feasible space illustrated by Figure 4. At every time step, the following min–max problem is solved:

$$Y^{risk} = \arg\min_{Y}(-\Phi(Y)) \tag{46}$$

where $\Phi(Y)$ is the voltage undershooting/overshooting corresponding to input values of Y, as defined in Section 5.1, subject to:

$$\widehat{Y}(1) - \delta_{\text{PV}} \leqslant Y(1) \leqslant \widehat{Y}(1) + \delta_{\text{PV}} \tag{47}$$
$$\widehat{Y}(2) - \delta_c \leqslant Y(2) \leqslant \widehat{Y}(2) + \delta_c \tag{48}$$
$$\widehat{Y}(3) - \delta_w \leqslant Y(3) \leqslant \widehat{Y}(3) + \delta_w \tag{49}$$

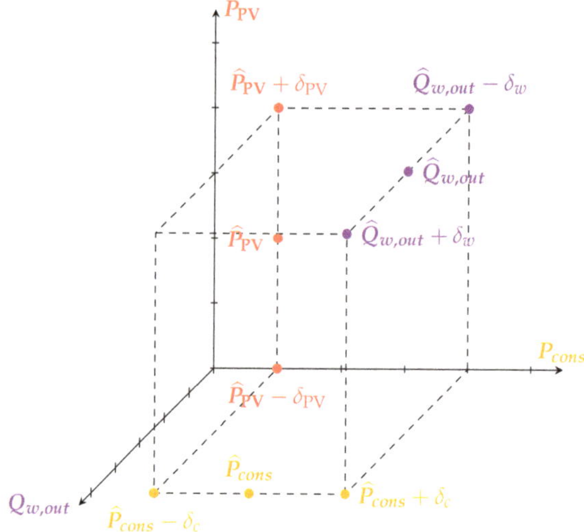

Figure 4. Feasible space of the min–max problem, defined by the confidence intervals of one-step-ahead forecasts of grid load, water demand, and PV power generation. The time indices are removed to avoid cluttering the illustration. All quantities in this figure correspond to values in the next time slot.

5. Results and Analysis

5.1. Evaluation Metrics

Various evaluation metrics used throughout this paper are defined hereinafter. Let H_t be the number of time slots in the simulation period.

1. Final value of the objective function: its square root ($\sqrt{f_{obj,final}}$) represents the cumulative gap between power supply and demand within the power distribution grid during the simulated week.
2. Computational complexity κ: it is quantified by the mean number of objective function evaluations per window, weighted by its size. The number of objective function evaluations is provided as an output argument of the optimisation function "fmincon" in Matlab.
3. Mean deviation from the forecasted values: let $\bar{P}_{PV}$, $\bar{P}_{cons}$, and $\bar{Q}_{w,out}$ be vectors grouping one-step-ahead forecasts (herein, 10-min forecasts) of PV power generation, grid load, and water demand, respectively, during the simulation period. This evaluation metric represents the mean deviation of stochastic input values from the ones forecasted at a one-step-ahead forecast horizon. $\Omega_{P_{PV}}$, which is the mean deviation of PV power generation values (P_{PV}) from the ones forecasted at a one-step-ahead forecast horizon (in W), is given by:

$$\Omega_{P_{PV}} = \frac{\sum_{t=0}^{H_t-1} \bar{P}_{PV}(t+1) - P_{PV}(t+1)}{H_t} \tag{50}$$

$\Omega_{P_{cons}}$, which is the mean deviation of grid load values (P_{cons}) from the ones forecasted at a one-step-ahead forecast horizon (in W), is given by:

$$\Omega_{P_{cons}} = \frac{\sum_{t=0}^{H_t-1} \bar{P}_{cons}(t+1) - P_{cons}(t+1)}{H_t} \tag{51}$$

$\Omega_{Q_{w,out}}$, which is the mean deviation of water demand values ($Q_{w,out}$) from the ones forecasted at a one-step-ahead forecast horizon (in $m^3\,h^{-1}$), is given by:

$$\Omega_{Q_{w,out}} = \frac{\sum_{t=0}^{H_t-1} \bar{Q}_{w,out}(t+1) - Q_{w,out}(t+1)}{H_t} \tag{52}$$

4. Instances of voltage overshooting ν: in cases where the main optimisation problem has no feasible solution, overvoltage or undervoltage may occur in the power distribution grid. These are the instances that the present paper studies and attempts to reduce. This metric records the percentage of these instances during the simulation period.
5. Surface area of voltage overshooting Φ: this metric is complementary to the number of instances of voltage constraint violation. It represents the total surface area of voltage overshooting past the prescribed lower and upper voltage levels in the power distribution grid, during the simulated period. It is measured in volts and is calculated as follows:

$$\Phi(Y) = \sum_{s=1}^{N} \left(\sum_{t=1}^{H_t} \max(U_s(t) - U_{min}, U_{max} - U_s(t), 0) \right) \tag{53}$$

where N is the number of nodes in the power distribution grid, U_{min} is the lower voltage threshold, U_{max} is the upper voltage threshold, and U_s, where $s \in \{1, \ldots, N\}$, are voltage values in various nodes of the power distribution grid. Let $\Psi \in \mathbb{R}^{H_p \times 3}$ be the matrix containing the measured stochastic inputs of the control strategy, over the simulation period, defined as follows:

$$\Psi = \begin{bmatrix} P_{PV} & P_{cons} & Q_{w,out} \end{bmatrix} \tag{54}$$

Note that voltage values across the power distribution grid are linked to the values of Ψ through Kirchhoff laws.

6. Average voltage overshooting per time step ϕ: it corresponds to the mean of maximum voltage overshooting over the number of instances at which overshooting is observed during the simulation period. It is defined as follows:

$$\phi(\mathbf{Y}) = 100\frac{\Phi_{max}(\mathbf{Y})}{v \cdot H_t} \tag{55}$$

with

$$\Phi_{max}(\mathbf{Y}) = \sum_{t=1}^{H_t} \max_{s \in \{1,...,N\}} (\mathbf{U}_s(t)) \tag{56}$$

where values of $\mathbf{Y}$ and $\mathbf{U}_s$ are related through Kirchhoff laws.

5.2. Impact of Forecasting Errors on MPC Performance

In this section, the impact of forecasting errors on the MPC strategy's performance is studied. The control scheme is tested over the week in April presented in the previous section and for sliding window sizes ranging from 1 to 24 h. Intraday GPR forecasts of grid load, water demand, and PV power generation (inferred from GHI forecasts), acquired as explained in Section 3, are used to run these simulations.

The performance of the predictive controller fed with GPR forecasts is evaluated in comparison with a controller fed with measurements, i.e., a case where no forecasting errors are made. This comparison will focus on three main aspects of the scheme's performance: the power supply/demand gap $\sqrt{f_{obj,,final}}$ (Figure 5), computational cost κ (Figure 6), and surface area of voltage overshooting Φ (Figure 7).

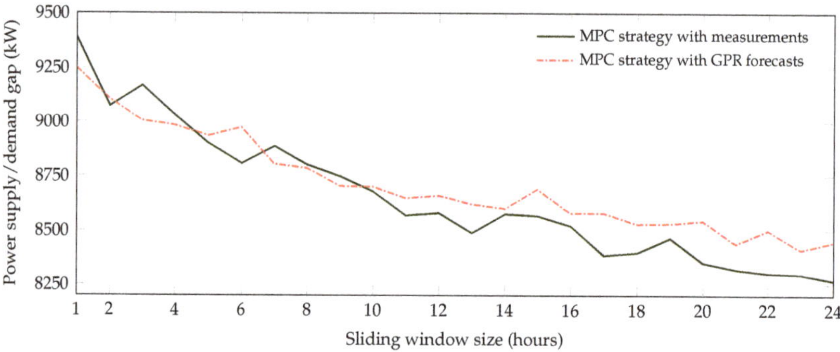

Figure 5. The cumulative power supply/demand gap within the power distribution grid, per sliding window size. The gap before optimisation is 10,035 kW.

In Figure 5, the cumulative power supply/demand gap given by the MPC scheme over the considered week is displayed per sliding window size. Though these values degrade as the sliding window size increases in both cases, the ones given by the MPC scheme when it uses GPR forecasts are not significantly degraded with respect to those given by the MPC scheme that uses measurements. In fact, the maximum difference between the two curves is 198.39 kW, obtained for the 22-h sliding window, which constitutes only 1.98% of the initial value (10,035 kW).

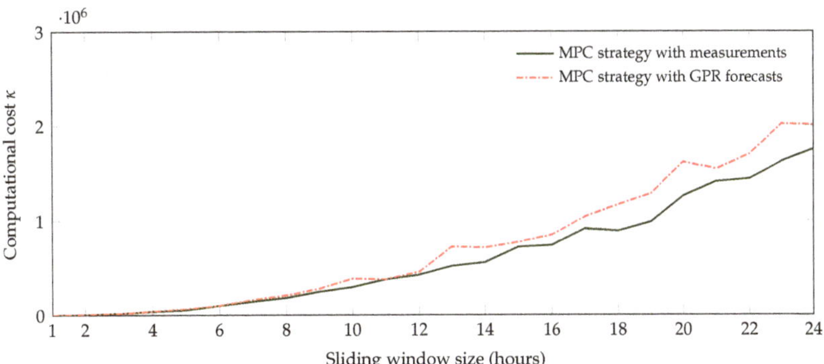

Figure 6. Computational complexity, measured as the mean number of function evaluations per sliding window weighted by its size.

The computational cost, presented in Figure 6, steadily grows in both cases as the sliding window size does. It is different for the scheme that uses GPR forecasts, which is to be expected since the updated forecasts displace the optimisation problem's starting point with respect to the previous time step. That being said, the increase in computational cost due to the use of forecasts remains subdued. For window sizes between 1 and 10 h, its average value is a 12.3% increase from the scheme using measurements. For all window sizes up to 24 h, this average is evaluated at 16.4%.

Figure 7 displays the surface area of voltage overshooting, the initial value of which is 4371.4 kV. For sliding window sizes up to 3 h, the MPC scheme is unable to reduce voltage overshooting and actually makes matters considerably worse. Starting from a 4-h sliding window, voltage overshooting given by the MPC scheme decreases significantly from the initial value, in both the case where the scheme uses measurements and the one where it uses GPR forecasts. Then, for larger window sizes, it quickly stabilizes around the same level. As of the 4 h window size, the MPC strategy effectively eliminates more than 50% of voltage overshooting.

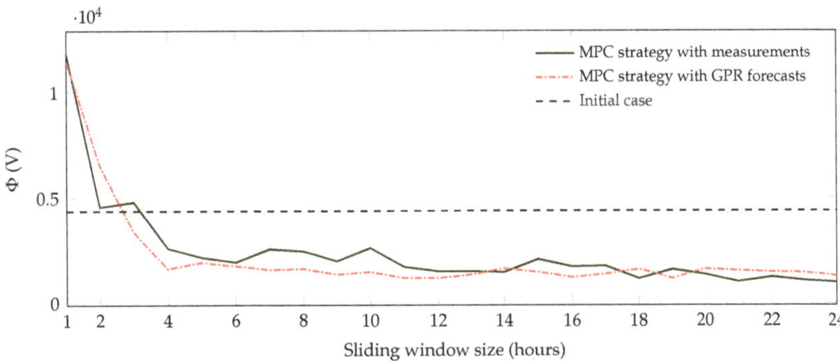

Figure 7. The total surface area of voltage overshooting per sliding window size.

When taking into account the fact that the forecasting errors are at their lowest for short forecast horizons (inferior to 3 h) and rapidly grow for longer horizons, it becomes apparent that the accuracy of forecasts for these short horizons is not enough to guarantee better management of voltage fluctuations on its own. In reality, the availability of forecasts over a longer forecast horizon is pivotal to better equip the MPC scheme to anticipate emerging voltage overshooting and work to prevent it. In light of this observation, it is recommended, for the purposes of this study, to prioritise reducing forecasts' error rates for forecast horizons up to several hours, rather than only focusing on short horizons.

Figure 8 displays the percentages of time steps during the simulated week where overshooting is observed, with respect to the size of the sliding window. As of the 2-h window, this percentage is significantly lower than the initial one (23.71%) for both MPC schemes, the one using measurements and the one using GPR forecasts. It quickly stabilizes at roughly 7% for both cases and reaches a minimum of 3.67% for the former and of 5.16% for the latter, both corresponding to the 24 h window.

Figure 9 displays the average voltage overshooting per time step (ϕ), with respect to the sliding window size. These values are lower than the initial value (18.29%) for both schemes, with the values corresponding to the MPC scheme using GPR forecasts being slightly lower than the ones corresponding to the MPC scheme using measurements. It is interesting to note that despite the percentage of overshooting occurrences (ν) being significantly higher for small windows than for larger ones, values of ϕ remain at roughly the same level regardless of sliding window size. Their average is 11.6 V per time step for the scheme using measurements and 10.7 V per time step for the one using GPR forecasts. This means that, for large window sizes, overshooting incidents are less frequent, but have a higher amplitude than for small ones.

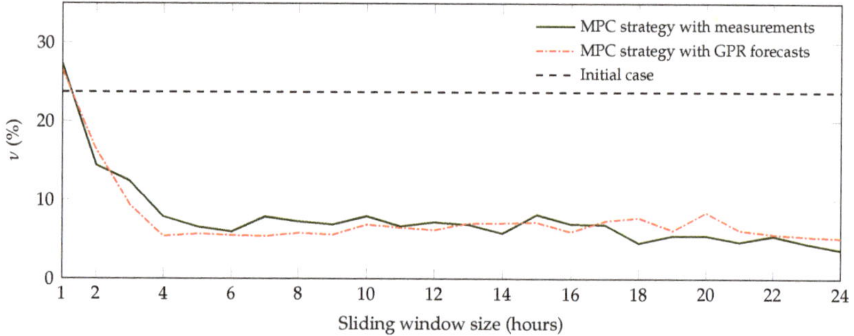

Figure 8. Percentage of time steps where an overshooting is observed, with respect to the sliding window size used by the MPC scheme.

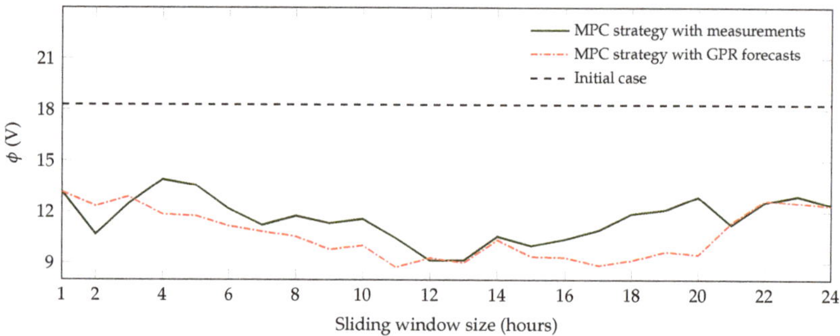

Figure 9. Average voltage overshooting per time step, with respect to the sliding window size used by the MPC scheme.

On another note, voltage overshooting is not remarkably impacted by the use of GPR forecasts as opposed to the case where measurements are used. The two sets of values, whether in terms of cumulative voltage overshooting (Figure 7) or percentage-wise (Figure 8), behave similarly and remain roughly at the same level. These results point to the MPC strategy proposed herein being inherently resilient to forecasting errors of PV power generation, grid load, and water demand. This is thanks to the closed-loop structure of model-based predictive control, which allows course-correction as new information comes in at every time step.

Although the MPC strategy succeeds in reducing voltage overshooting with respect to the initial case, it does not completely eliminate it. In order to further enhance the strategy's performance, a complementary module can be added upstream of the main optimisation problem upon which the predictive controller is based. This module attempts to limit overshooting due to erroneous GPR forecasts of grid load, water demand, and PV power generation.

5.3. Contribution of the Worst-Case Approach

It should be noted here that the main optimisation problem responsible for balancing power supply and demand in the power distribution grid, as constructed in Section 4.2, prioritises constraints so that the volume ones are always upheld. In other words, in cases where no feasible solution is found, voltage constraints are relaxed in order to guarantee that biogas volume constraints and water volume constraints are always upheld. For this reason, hereinafter, only voltage constraint violation is examined to evaluate the efficiency of the proposed min–max problem in enhancing the control strategy's robustness to forecasting errors.

In this section, the amended predictive control strategy as explained above is implemented, and its results are presented and discussed in comparison with three other cases:

- Case 1. The initial case where no optimisation is carried out. The biogas plant has a constant power generation output. The water tower is subject to an ON/OFF controller, which activates its pump when a low-level sensor is triggered and deactivates it when a high-level sensor is triggered;
- Case 2. The predictive control strategy described in Section 4, with GPR forecasts of the PV power generation, grid load, and water demand used;
- Case 3. The amended control strategy proposed in this paper, based on the addition of the min–max problem to anticipate the worst-case scenario within the forecasts' confidence intervals in terms of constraint violation.

For each case, a simulation is run over a week in April. This period is selected because it presents high PV power generation and therefore demonstrates significant voltage overshooting. Two sizes of sliding windows are used for the MPC scheme in the results that are presented hereinafter: a 4 h window and a 10 h window. These two sliding window sizes are chosen to examine the difference in effects of the min–max problem on the MPC strategy's performance for both short sliding windows and long ones. The evaluation metrics of the MPC scheme with both sliding window sizes are assembled in Table 1.

Table 1. Assessment of the min–max problem's contribution to the control strategy's robustness to forecasting errors, for a week in April. See below (Section 5.3) for details about the three cases. For Case 2 and Case 3, 4 h and 10 h sliding windows are considered.

Evaluation Metric	Case 1	Case 2		Case 3	
		4-h	10-h	4-h	10-h
$\sqrt{f_{obj,final}}$ (kW)	10,035	8984	8700	8966	8648
κ (–)	–	40,872	385,320	39,700	358,990
$\Omega_{P_{PV}^{risk}}$ (kW)	–	–	–	16.26	19.41
$\Omega_{P_{cons}^{risk}}$ (kW)	–	–	–	5.56	6.73
$\Omega_{Q_{w,out}^{risk}}$ (m^3 h^{-1})	–	–	–	-4.24×10^{-15}	-6.41×10^{-16}
ν (%)	23.71	5.36	6.85	4.96	6.35
Φ (kV)	4371.4	1632.7	1464.6	1464.5	1176.8

When examining the inner-workings of the min–max problem, it can be deduced that there are noteworthy deviations between forecasted values of PV power generation and grid load and the ones corresponding to worst-case scenarios. For the considered week, for the tested MPC windows of 4 h and 10 h, the deviation of worst-case scenario PV power generation values from the forecasted values ($\Omega_{P_{\text{pv}}}$) represents, on average, 7% and 9% of the data's mean, respectively. When it comes to the grid load, the deviation of worst-case scenario values from forecasted ones $\Omega_{P_{cons}}$ is less notable. For the tested MPC windows of 4 h and 10 h, it represents 3% an 2% of the data's mean, respectively.

Having said that, forecasted values of water demand and the ones corresponding to worst-case scenarios are virtually identical ($\Omega_{Q_{w,out}}$ is virtually null). This observation reaffirms the presumption that water demand, and by extension water levels in the water tower's storage tank, have no direct impact on voltage fluctuations in the power distribution grid. Their influence resides solely in determining the water tower's capacity in absorbing excess power off the grid at a given time, which can be properly foreseen through proper dimensioning of the storage tank.

The instances of voltage overshooting decrease steadily from Case 1 through Case 3. In fact, the amended MPC scheme with the min–max problem (Case 3) brings down their percentage (ν) to 4.96% and 6.35% for the 4 h window and the 10 h window, respectively, from an initial value of 23.71%. The total surface area of voltage overshooting Φ is also considerably reduced from the initial value. It brought down to 1464.5 kV and 1176.8 kV for the 4 h window and the 10 h window, respectively, from an initial value of 4371.4 kV.

The gain procured through the addition of the min–max problem to the MPC scheme is deduced by comparing the metrics of Case 2 and Case 3. As it happens, for the 4 h sliding window, voltage overshooting is further decreased from Case 2 to Case 3 by 168.2 kV, which amounts to 3.8% of the total surface area of voltage overshooting in the initial case. Percentage-wise, this decrease corresponds to 0.4% of the initial instances of overshooting.

For the 10 h sliding window, voltage overshooting is decreased from Case 2 to Case 3 by 287.8 kV, which amounts to 6.6% of the total surface area of voltage overshooting in the initial case. Percentage-wise, this decrease corresponds to 0.5% of the initial instances of overshooting. The small fraction of the eliminated instances of overshooting with respect to the corresponding percentage of reduced surface area suggests that the min–max problem is particularly apt in eliminating major overshooting incidents. Table 1 reveals that the drop in the total surface area of voltage overshooting observed between the scheme using a 4 h window and the one using a 10 h window is accompanied by an increase in the percentage of instances of overvoltage. Figure 10 illustrates the extrema of voltage fluctuations within the power distribution grid for the standard MPC strategy and the one using a min-max problem, for both sliding window sizes (4 h and 10 h). Voltage overshooting is considerably reduced in both cases with respect to the initial case. Voltage values mostly remain within the acceptable voltage bounds and veer closer to the nominal value (230 V). Unfortunately, this is achieved at the expense of the smoothness of the voltage curves. A possible solution to this issue could be the addition of a regularisation term to the objective function in order to penalise high-frequency voltage fluctuations. That being said, several voltage fluctuations are eliminated thanks to the addition of the min–max problem to the control strategy. This is especially noticeable for the MPC scheme using a 10 h window where notable overshooting incidents are removed during midday of 16 April and 18 April.

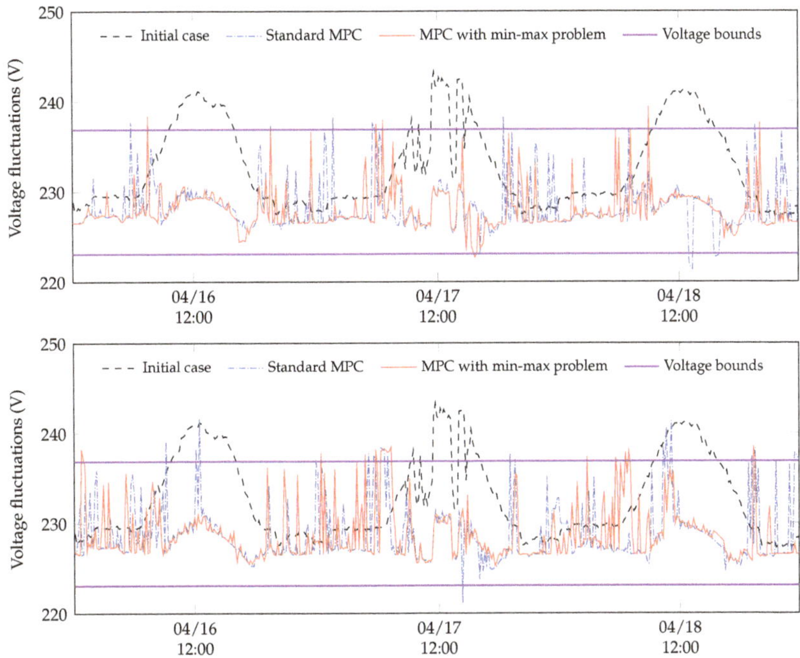

Figure 10. Extrema of voltage fluctuations within the power distribution grid for the standard MPC strategy compared to the one using a min–max problem and to the initial case, displayed for a 4 h sliding window (**top**) and a 10 h sliding window (**bottom**).

The addition of the min–max problem does not introduce additional computational burden to the control strategy. In fact, the computational complexity (measured by κ) decreases from Case 2 to Case 3 by 2.9% for the scheme using a 4 h window and by 6.8% for the scheme using a 10 h window. Figure 11 displays the evolution of the gap between power supply and demand for MPC schemes with and without the min–max problem. It is clear that neither scheme succeeds in reducing this gap significantly. However, they trim the peak occurring everyday around noon, due to the peak in PV power generation. This trimming effect is more visible for the scheme using a 10 h window than the one using a 4 h window. This is reflected in the the final values of the objective function. Though reduced from the initial case, they change very little between Case 2 (MPC using GPR forecasts) and Case 3 (MPC using GPR forecasts and the min–max problem). This suggests that the min–max problem does not provoke any degradation to the MPC strategy's ability to reduce the gap between supply and demand in the power distribution grid.

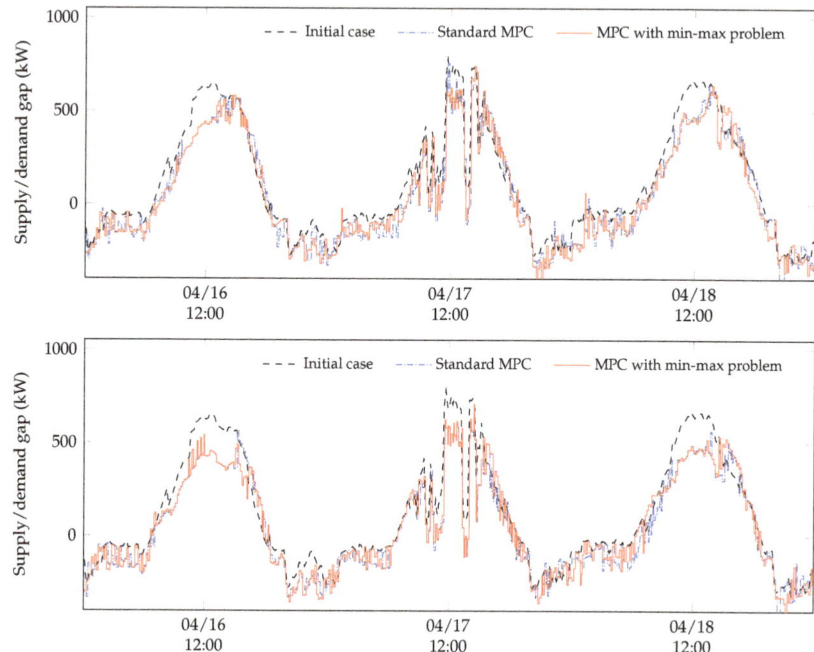

Figure 11. Gap between power supply and demand within the power distribution grid for the standard MPC strategy compared to the one using a min–max problem and to the initial case, displayed for a 4 h sliding window (**top**) and a 10 h sliding window (**bottom**).

In the case study considered here, it can be seen that the MPC strategy's reduction of the gap between supply and demand remains humble. This may be traced back to the dimensioning of the flexible assets, especially the dimensioning of these assets' storage units. One could argue that the power generation capacity of the chosen biogas plant and the power demand capacity of the chosen water tower are too small to have any meaningful impact on the reduction of the supply/demand gap within the power distribution grid. This observation further illustrates the importance of optimal dimensioning of flexible assets in order for the smart management scheme to yield efficient results.

5.4. Discussion

The introduction of the worst-case scenario approach, detailed in Section 4.3, is inspired by previous works on min–max optimisation for uncertain nonlinear systems under constraints, which is by definition a conservative risk aversion technique. It is chosen for its relative ease of implementation and is used in this paper to complement the previously developed predictive control strategy in order to anticipate the worst-case scenario, in terms of stochastic input values, and minimise resulting voltage overshooting. The conservativeness of the technique is lessened by its containment to the following time step instead of the entire forecast horizon. This is done primarily to reduce the computational burden added by the min–max problem, in light of the real-time aspect of the application at hand. Nevertheless, the merit of extending the min–max problem to the entire forecast horizon and a quantification of its added cost are valid questions worth investigating in a follow-up to this work. It is worth mentioning that, on top of reducing voltage overshooting, the min–max problem has virtually no effect on the reduction of the gap between power supply and demand.

It is clear that both the growing power demand and the deployment of distributed generation within power distribution grids are not slowing down in the near future. Therefore, configurations like the one studied herein will likely materialise in the upcoming years, accompanied by emerging constraints such as the ones observed in this study. This goes to show the pertinence of studies such as this one in order to prepare for this inevitability.

The damper on the control scheme's performance in the configuration studied in this paper can be traced back, at least partially, to the flexible assets' dimensioning and their suitability to the task. As grid load and PV power generation levels rapidly rise, the flexible assets' room for manoeuvring diminishes. Consequently, two critical questions transpire. The first question is that of optimal dimensioning of these flexible assets. In this study, the control strategy operates at the level of the MV/LV transformer. However, in an application with finer spatial granularity, this question also encompasses optimal placement of the assets within the grid. The second question is that of the assets nature and their suitability to the application.

The combination of a biogas plant and a water tower in this case study was selected in an attempt to utilise small-scale assets compatible with the rural setting of the "Smart Occitania" project and offer the possibility of deferment of the assets' operation. Having said that, the examination of this duo's potential reveals several flaws. To begin, the water tower's ON/OFF operation adds computational complexity to the optimisation problem and is therefore a handicap to the real-time aspect of the applications. Besides, it infers choppier setpoints, which not only worsen voltage fluctuations but also shorten the equipment's life expectancy.

This is especially problematic when the installation's latitude in terms of storage levels is limited. Furthermore, flexible assets need to be extensible in order to adapt their room for manoeuvring to the rapidly growing power demand and distributed generation levels within the grid with minimal cost. The assets considered herein are not easily extensible, particularly the biogas plant, whose energy source is based on a fairly delicate organic process.

In the case of model-based predictive control applications for power grids, choosing the time step is a pivotal task with no definitive answer. A compromise is always made between the high computational cost of this type of control scheme and the granularity of the model, which allows us to capture a maximum of phenomena occurring in the system. This type of control strategy is also dependent upon access to data, in real-time, which comes with its own set of technical issues. Solutions to these issues are starting to come together through the maturing of advanced metering infrastructures in recent years.

The 10 min time step considered in this work is very much an instance of the aforementioned compromise. It allows the necessary computations of both the forecast module and the optimisation problem to run their course. However, it limits the strategy's visibility into the high-dynamics of power grids and thus makes it impossible to intervene between two sampling times. This type of strategy can therefore be seen as an-upper level control scheme, to be coupled with longer-term planning strategies and lower-level operation methods that have the capacity to react to rapid electrical phenomena, namely methods that fall under the umbrella of electrotechnical engineering.

6. Conclusions and Prospects

The work presented in this manuscript falls within the scope of the "Smart Occitania" project, whose goal is to demonstrate the feasibility of the smart grid concept for rural and suburban low-voltage power distribution grids. To this end, a simulated case study is constructed, based on data collected during the project's run and made available by ENEDIS, in order to elaborate a predictive control strategy for more efficient management of power flows within a power distribution grid with prolific levels of distributed generation, namely PV power generation.

The premise of the proposed strategy is to use a model-based predictive controller to optimise setpoints of flexible assets present in the power distribution grid, in order to reduce the gap between supply and demand. This optimisation is constrained by pre-defined acceptable voltage margins, in addition to the assets' operational restrictions.

A forecast module is constructed using Gaussian process regression and provides intraday forecasts of grid load, water demand, and GHI from which PV power generation is inferred. The quality of the forecasts decreases as the forecast horizon grows longer, but quickly stabilizes around a constant error value. The GPR models also provide confidence intervals associated with the forecasts. Herein, the computed confidence intervals correspond to a probability of 95% of containing the measurements.

An evaluation is carried out of the MPC strategy's resilience to forecasting errors, and the induced errors are quantified. Results show that the predictive control strategy is inherently resilient to forecasting errors as the final objective function value varies little between the case where measurements are used and the one where GPR forecasts are used.

Finally, a min–max problem is added upstream of the main optimisation problem. Its purpose is to anticipate and minimise the voltage overshooting resulting from forecasting errors. In this min–max problem, the feasible space defined by the confidence intervals of the forecasts is searched, in order to determine the worst-case scenario in terms of constraint violation, over the next time step. Then, the main optimisation problem incorporates this information into its decision-making process. Results show that these incidents are indeed reduced thanks to the min–max problem, both in terms of frequency of their occurrence and the total surface area of overshooting.

There are several axes along which the present work can make headway. To begin, the optimisation problem could be modified to take into account the implementability of the flexible assets' setpoints. For example, this could take the form of a multi-objective optimisation that balances out sometimes-conflicting goals.

In addition, the min–max problem integrated into the predictive control strategy in order to improve the scheme's resilience to forecasting errors, can be extended from focusing on the next time step to span the entire forecast horizon. The length of the min–max problem's forecast horizon would be another parameter to be optimised. This will inevitably increase the computational burden of the control scheme, but should also enhance its performance through a better anticipation of issues that may arise along the forecast horizon, especially in light of the degradation of the forecasts' quality as the algorithm advances into the forecast horizon.

Author Contributions: Conceptualization, N.D. and J.E.; methodology, N.D. and J.E.; formal analysis, N.D. and J.E.; investigation, N.D., J.E., S.T. and S.G.; writing—original draft preparation, N.D.; writing—review and editing, J.E., S.T. and S.G.; supervision, J.E., S.T. and S.G.; project administration, S.G.; funding acquisition, S.G. All authors have read and agreed to the published version of the manuscript.

Funding: This research was funded by the French agency for ecological transition (ADEME).

Institutional Review Board Statement: Not applicable.

Informed Consent Statement: Not applicable.

Data Availability Statement: The data presented in this study are available on request from the corresponding author.

Acknowledgments: The authors would like to thank the French agency for ecological transition for its financial support. They also thank all the academic and industrial entities involved in the "Smart Occitania" project for their contribution to this work.

Conflicts of Interest: The authors declare no conflict of interest.

Abbreviations

The following abbreviations are used in this manuscript:

ADEME	French agency for ecological transition
ANN	Artificial neural network

CWC	Coverage width-based criterion
DSM	Demand side management
GHI	Global horizontal irradiance
GP	Gaussian process
GPR	Gaussian process regression
LHV	Low heating value of the stored biogas
LSTM	Long short term memory
LV	Low voltage
MAS	Multi-agent systems
MINLP	Mixed-integer nonlinear programming
MPC	Model-based predictive control
MV/LV	Medium-voltage/low-voltage
nRMSE	Normalized root mean square error
OPF	Optimal power flow
PICP	Prediction interval coverage probability
PINAW	Prediction interval normalized average width
PROMES-CNRS	Processes, materials and solar energy
PV	Solar photovoltaics
RQ	Rational quadratic kernel (Gaussian process regression)
SE	Squared exponential kernel (Gaussian process regression)

References

1. Direction Technique ENEDIS. *Principes D'étude et de Développement du Réseau Pour le Raccordement des Clients Consommateurs et Producteurs BT*; Technical Report; ENEDIS: Paris La Défense, France, 2018.
2. Abdi, H.; Beigvand, S.D.; La Scala, M. A review of optimal power flow studies applied to smart grids and microgrids. *Renew. Sustain. Energy Rev.* **2017**, *71*, 742–766. [CrossRef]
3. Syranidis, K.; Robinius, M.; Stolten, D. Control techniques and the modeling of electrical power flow across transmission networks. *Renew. Sustain. Energy Rev.* **2018**, *82*, 3452–3467. [CrossRef]
4. De Oliveira-De Jesus, P.M.; Henggeler Antunes, C. A detailed network model for distribution systems with high penetration of renewable generation sources. *Electr. Power Syst. Res.* **2018**, *161*, 152–166. [CrossRef]
5. Joshi, K.; Pindoriya, N. Advances in Distribution System Analysis with Distributed Resources: Survey with a Case Study. *Sustain. Energy Grids Netw.* **2018**, *15*, 86–100. [CrossRef]
6. Frank, S.; Rebennack, S. An introduction to optimal power flow: Theory, formulation, and examples. *IIE Trans.* **2016**, *48*, 1172–1197. [CrossRef]
7. Riffonneau, Y.; Bacha, S.; Barruel, F.; Ploix, S. Optimal Power Flow Management for Grid Connected PV Systems With Batteries. *IEEE Trans. Sustain. Energy* **2011**, *2*, 309–320. [CrossRef]
8. Strbac, G. Demand side management: Benefits and challenges. *Energy Policy* **2008**, *36*, 4419–4426. [CrossRef]
9. Palensky, P.; Dietrich, D. Demand side management: Demand response, intelligent energy systems, and smart loads. *IEEE Trans. Ind. Inf.* **2011**, *7*, 381–388. [CrossRef]
10. Meyabadi, A.F.; Deihimi, M.H. A review of demand-side management: Reconsidering theoretical framework. *Renew. Sustain. Energy Rev.* **2017**, *80*, 367–379. [CrossRef]
11. Balijepalli, V.M.; Pradhan, V.; Khaparde, S.A.; Shereef, R. Review of Demand Response under Smart Grid Paradigm. In Proceedings of the ISGT2011-India, Kollam, India, 1 December–3 December 2011, pp. 236–243.
12. Aghaei, J.; Alizadeh, M.I. Demand response in smart electricity grids equipped with renewable energy sources: A review. *Renew. Sustain. Energy Rev.* **2013**, *18*, 64–72. [CrossRef]
13. Siano, P. Demand response and smart grids—A survey. *Renew. Sustain. Energy Rev.* **2014**, *30*, 461–478. [CrossRef]
14. McArthur, S.D.J.; Davidson, E.M.; Catterson, V.M.; Dimeas, A.L.; Hatziargyriou, N.D.; Ponci, F.; Funabashi, T. Multi-Agent Systems for Power Engineering Applications—Part I: Concepts, Approaches, and Technical Challenges. *IEEE Trans. Power Syst.* **2007**, *22*, 1743–1752. [CrossRef]
15. McArthur, S.D.J.; Davidson, E.M.; Catterson, V.M.; Dimeas, A.L.; Hatziargyriou, N.D.; Ponci, F.; Funabashi, T. Multi-Agent Systems for Power Engineering Applications – Part II: Technologies, Standards and Tools for Building Multi-Agent Systems. *IEEE Trans. Power Syst.* **2007**, *22*, 1753–1759. [CrossRef]
16. Mocci, S.; Natale, N.; Pilo, F.; Ruggeri, S. Demand side integration in LV smart grids with multi-agent control system. *Electr. Power Syst. Res.* **2015**, *125*, 23–33. [CrossRef]
17. You, S.; Segerberg, H. Integration of 100% micro-distributed energy resources in the low voltage distribution network: A Danish case study. *Appl. Therm. Eng.* **2014**, *71*, 797–808. [CrossRef]
18. Haque, A.N.M.M.; Nguyen, P.H.; Vo, T.H.; Bliek, F.W. Agent-based unified approach for thermal and voltage constraint management in LV distribution network. *Electr. Power Syst. Res.* **2017**, *143*, 462–473. [CrossRef]

19. Dkhili, N.; Eynard, J.; Thil, S.; Grieu, S. A survey of modelling and smart management tools for power grids with prolific distributed generation. *Sustain. Energy Grids Netw.* **2020**, *21*, 100284. [CrossRef]
20. Dkhili, N.; Salas, D.; Eynard, J.; Thil, S.; Grieu, S. Innovative application of model-based predictive control for low-voltage power distribution grids with significant distributed generation. *Energies* **2021**, *14*, 1773. [CrossRef]
21. Pesaran, M.H.A.; Huy, P.D.; Ramachandaramurthy, V.K. A review of the optimal allocation of distributed generation: Objectives, constraints, methods, and algorithms. *Renew. Sustain. Energy Rev.* **2017**, *75*, 293–312. [CrossRef]
22. Sugihara, H.; Yokoyama, K.; Saeki, O.; Tsuji, K.; Funaki, T. Economic and efficient voltage management using customer-owned energy storage systems in a distribution network with high penetration of photovoltaic systems. *IEEE Trans. Power Syst.* **2013**, *28*, 102–111. [CrossRef]
23. Karimyan, P.; Gharehpetian, G.B.; Abedi, M.R.; Gavili, A. Long term scheduling for optimal allocation and sizing of DG unit considering load variations and DG type. *Int. J. Electr. Power Energy Syst.* **2014**, *54*, 277–287. [CrossRef]
24. Bruni, G.; Cordiner, S.; Mulone, V.; Rocco, V.; Spagnolo, F. A study on the energy management in domestic micro-grids based on model predictive control strategies. *Energy Convers. Manag.* **2015**, *102*, 50–58. [CrossRef]
25. Prodan, I.; Zio, E. A model predictive control framework for reliable microgrid energy management. *Int. J. Electr. Power Energy Syst.* **2014**, *61*, 399–409. [CrossRef]
26. Vazquez, S.; Leon, J.I.; Franquelo, L.G.; Rodriguez, J.; Young, H.A.; Marquez, A.; Zanchetta, P. Model predictive control: A review of its applications in power electronics. *IEEE Ind. Electron. Mag.* **2014**, *8*, 16–31. [CrossRef]
27. Tøndel, P.; Johansen, T.A. Complexity reduction in explicit linear model predictive control. *IFAC Proc. Vol.* **2002**, *35*, 189–194. [CrossRef]
28. Belotti, P.; Kirches, C.; Leyffer, S.; Linderoth, J.; Luedtke, J.; Mahajan, A. Mixed-integer nonlinear optimization. *Acta Numer.* **2013**, *22*, 1–131. [CrossRef]
29. Burer, S.; Letchford, A.N. Non-convex mixed-integer nonlinear programming: A survey. *Surv. Oper. Res. Manag. Sci.* **2012**, *17*, 97–106. [CrossRef]
30. Lee, J.; Leyffer, S. *Mixed Integer Nonlinear Programming*; Springer: New York, NY, USA, 2012.
31. Nocedal, J.; Wright, S.J. *Numerical Optimization*; Springer: New York, NY, USA, 2006.
32. Tawarmalani, M.; Sahinidis, N.V.; Sahinidis, N. *Convexification and Global Optimization in Continuous and Mixed-Integer Nonlinear Programming: Theory, Algorithms, Software, and Applications*; Springer Science & Business Media: New York, NY, USA, 2002; Volume 65.
33. Sahinidis, N. Mixed-integer nonlinear programming 2018. *Optim. Eng.* **2019**, *20*, 301–306. [CrossRef]
34. Trespalacios, F.; Grossmann, I. Review of mixed-integer nonlinear and generalized disjunctive programming methods. *Chem. Ing. Tech.* **2014**, *86*, 991–1012. [CrossRef]
35. Bemporad, A.; Morari, M. Robust model predictive control: A survey. In *Robustness in Identification and Control*; Springer: Berlin/Heidelberg, Germany, 1999; pp. 207–226.
36. Mayne, D.Q.; Seron, M.M.; Raković, S. Robust model predictive control of constrained linear systems with bounded disturbances. *Automatica* **2005**, *41*, 219–224. [CrossRef]
37. Di Cairano, S.; Bernardini, D.; Bemporad, A.; Kolmanovsky, I.V. Stochastic MPC with learning for driver-predictive vehicle control and its application to HEV energy management. *IEEE Trans. Control. Syst. Technol.* **2013**, *22*, 1018–1031. [CrossRef]
38. Köhler, J.; Soloperto, R.; Müller, M.A.; Allgöwer, F. A computationally efficient robust model predictive control framework for uncertain nonlinear systems. *arXiv* **2019**, arXiv:1910.12081.
39. Fioriti, D.; Poli, D. A novel stochastic method to dispatch microgrids using Monte Carlo scenarios. *Electr. Power Syst. Res.* **2019**, *175*, 105896. [CrossRef]
40. Lucia, S.; Finkler, T.; Basak, D.; Engell, S. A new robust NMPC scheme and its application to a semi-batch reactor example. *IFAC Proc. Vol.* **2012**, *45*, 69–74. [CrossRef]
41. Lucia, S.; Andersson, J.A.; Brandt, H.; Diehl, M.; Engell, S. Handling uncertainty in economic nonlinear model predictive control: A comparative case study. *J. Process. Control.* **2014**, *24*, 1247–1259. [CrossRef]
42. Raimondo, D.M.; Limon, D.; Lazar, M.; Magni, L.; ndez Camacho, E.F. Min-max model predictive control of nonlinear systems: A unifying overview on stability. *Eur. J. Control.* **2009**, *15*, 5–21. [CrossRef]
43. Limón, D.; Alamo, T.; Salas, F.; Camacho, E.F. Input to state stability of min–max MPC controllers for nonlinear systems with bounded uncertainties. *Automatica* **2006**, *42*, 797–803. [CrossRef]
44. Skoplaki, E.; Palyvos, J.A. On the temperature dependence of photovoltaic module electrical performance: A review of efficiency/power correlations. *Sol. Energy* **2009**, *83*, 614–624. [CrossRef]
45. Jakhrani, A.Q.; Othman, A.K.; Rigit, A.R.H.; Samo, S.R. Comparison of Solar Photovoltaic Module Temperature Models. *World Appl. Sci. J. (Special Issue Food Environ.)* **2011**, *14*, 1–8.
46. Tolba, H.; Dkhili, N.; Nou, J.; Eynard, J.; Thil, S.; Grieu, S. GHI forecasting using Gaussian process regression: Kernel study. *IFAC-PapersOnLine* **2019**, *52*, 455–460. [CrossRef]
47. Gbémou, S.; Tolba, H.; Thil, S.; Grieu, S. Global horizontal irradiance forecasting using online sparse Gaussian process regression based on quasiperiodic kernels. In Proceedings of the 2019 IEEE International Conference on Environment and Electrical Engineering and 2019 IEEE Industrial and Commercial Power Systems Europe (EEEIC/I&CPS Europe), Genova, Italy, 4–10 June 2019.

48. Tolba, H.; Dkhili, N.; Nou, J.; Eynard, J.; Thil, S.; Grieu, S. Multi-Horizon Forecasting of Global Horizontal Irradiance Using Online Gaussian Process Regression: A Kernel Study. *Energies* **2020**, *13*, 4184. [CrossRef]
49. Rasmussen, C.E.; Williams, C.K.I. *Gaussian Processes for Machine Learning*; The MIT Press: Cambridge, MA, USA, 2006.
50. Khosravi, A.; Nahavandi, S.; Creighton, D.; Atiya, A.F. Lower upper bound estimation method for construction of neural network-based prediction intervals. *IEEE Trans. Neural Netw.* **2010**, *22*, 337–346. [CrossRef] [PubMed]

Article

Developing a Hybrid Optimization Algorithm for Optimal Allocation of Renewable DGs in Distribution Network

Ayman Awad [1], Hussein Abdel-Mawgoud [1], Salah Kamel [1], Abdalla A. Ibrahim [1] and Francisco Jurado [2,*]

[1] Department of Electrical Engineering, Faculty of Engineering, Aswan University, Aswan 81542, Egypt; engaymanelewy@gmail.com (A.A.); hussein.abdelmawgoud@yahoo.com (H.A.-M.); skamel@aswu.edu.eg (S.K.); abdalla.ibrahim@aswu.edu.eg (A.A.I.)

[2] Department of Electrical Engineering, University of Jaen, 23700 EPS Linares, Spain

* Correspondence: fjurado@ujaen.es

Abstract: Distributed generation (DG) is becoming a prominent key spot for research in recent years because it can be utilized in emergency/reserve plans for power systems and power quality improvement issues, besides its drastic impact on the environment as a greenhouse gas (GHG) reducer. For maximizing the benefits from such technology, it is crucial to identify the best size and location for DG that achieves the required goal of installing it. This paper presents an investigation of the optimized allocation of DG in different modes using a proposed hybrid technique, the tunicate swarm algorithm/sine-cosine algorithm (TSA/SCA). This investigation is performed on an IEEE-69 Radial Distribution System (RDS), where the impact of such allocation on the system is evaluated by NEPLAN software.

Keywords: power losses; power system optimization; PV curves; DG; TSA/SCA

Citation: Awad, A.;
Abdel-Mawgoud, H.; Kamel, S.;
Ibrahim, A.A.; Jurado, F. Developing
a Hybrid Optimization Algorithm for
Optimal Allocation of Renewable
DGs in Distribution Network. *Clean
Technol.* 2021, *3*, 409–423. https://
doi.org/10.3390/cleantechnol3020023

Academic Editor: Samuele Grillo

Received: 27 February 2021
Accepted: 7 April 2021
Published: 1 May 2021

Publisher's Note: MDPI stays neutral
with regard to jurisdictional claims in
published maps and institutional affiliations.

1. Introduction

Recently, the integration of distributed generation (DG) in distribution networks is becoming very popular to meet increases in system load [1–4]. Also, the world is interested in installing several types of DG, especially renewable sources in power grids, such as hydropower, biomass, photovoltaic (PV), and wind turbine [5–7] technologies. DG is a newly coined term that describes a technology that has a deep impact on power systems nowadays. A few years ago, distributed generation (DG)—or dispersed generation—started to arise in the world of power systems, aiming to exploit small-scale energy resources to utilize them in electric power generation instead of depending only on centralized large-scale power generation stations [1–4]. DG stations are characterized by being small-scale (usually less than 50 MW) and installed directly to the distribution power system, instead of the traditional transmission power system, enabling the facility that owns such a station to consume a part of the generated power, then export the surplus power [1,2].

Such a technology brings renewable energy resources into action, as most of the renewable-energy-dependent generation stations are small-scale stations, and of course, this will have a drastic effect on the environment. In other words, the more penetration of renewable-energy-dependent generation stations is achieved, the more reduction of greenhouse gasses (GHG) takes place [3–8].

Besides the environmental impact of distributed generation, it also has a remarkable impact on power-quality issues in distribution systems. It has a relieving effect on congested transmission and distribution systems due to its unique location: just beside the consumer! Such an advantage results in economic and environmental benefits by reducing the power losses of the system, as there is no need for additional transmission lines, while saving such losses results in a reduction of the GHG effect by about 1% [4].

On the other hand, distributed generation can be utilized for improving power quality in distribution systems, and even in transmission systems, as it has the ability to be con-

95

nected at different voltage levels, i.e., low, medium, and high voltage [1]. It can improve system reliability, maintain voltage stability, and also provide the system with reserve generated power for emergencies [1,5]. Moreover, such technology offers a remarkable contribution to economic investment in energy, inspiring researchers to introduce several techniques for the initiation and penetration of Distributed Energy Resources (DER) technologies and enabling a vision of such penetration regarding economic investment and cost reduction, focusing on biogas and hydrogen cells [9–12].

Nevertheless, identifying the perfect allocation and size of a distributed generator is a critical issue, as it increases the profit of installing it, which is the goal of installation. Obtaining the best location and size has a deep influence on the impact of installing DG. In [6], the optimal allocation of DG was identified with regards to the operations and investment costs of DG, where the utilized optimization approach tends to decrease the energy loss when the loads are time-varying by determining their generation capacity at different instants. Another optimization method is proposed in [7], regarding the minimization of power losses, where it is simplified so that it does not need excessive computational processes. In [8], a new optimization method, typically the virus colony search (VSC) algorithm was utilized, considering the reliability assessment of the distribution network, where the results are compared with several optimization methods. A hybrid method is proposed in [9] regarding loss minimization and voltage improvement, where the location of DG is identified using an empirical discrete metaheuristic (EDM), while the size is identified by the steepest descent method (SD).

TSA and SCA are efficient metaheuristic algorithms to solve difficult optimization problems [13–16]. Most metaheuristic algorithms have faced many challenges to determine a promising area of search space for their exploration, so the SCA algorithm is inserted into the TSA algorithm to improve the exploration phase of the TSA algorithm. There are several metaheuristic techniques used to obtain the optimal allocation of PVs and capacitors in RDS, such as whale optimization algorithm (WOA) [15], lightning search algorithm (LSA) [16], the backtracking search algorithm (BSA) [17], symbiotic organisms search (SOS) [18], crow search algorithm (CSA) [19], particle swarm optimization (PSO) algorithm [20], backtracking search optimization algorithm (BSOA) [21], firefly algorithm (FFA) [22], and the flower pollination algorithm (FPA) [23].

This paper introduces a new hybrid optimization approach wherein the tunicate swarm algorithm (TSA) is merged with the sine-cosine algorithm (SCA), resulting in a novel TSA/SCA hybrid approach. This new approach is used to identify the best size and location of DG in distribution, considering the minimization of system loss. The optimization process is performed in three scenarios: (1) the DG is producing real power only (P-type case), (2) the DG is producing reactive power only (Q-type case), and (3) the DG is generating both reactive and real power (PQ-type case). The performance of the RDS is evaluated after the optimization process using NEPLAN software.

The contributions of this paper are the (1) introduction of a new hybrid approach that consists of the TSA and SCA algorithms, (2) use of an efficient hybrid approach to determine the optimal planning of DG in RDS, and (3) study of the effect of integrating different types of DG in RDS.

The remainder of the paper is organized as follows: the presented problem is explained in Section 2; sensitivity is discussed in Section 3; Section 4 explains the presented algorithm and the obtained results. Section 5 discusses the conclusion.

2. Problem Formulation

Figure 1 shows the representation of two buses in a distribution system.

The system power flow is evaluated in a backwards direction by Equations (1) and (2) [24]:

$$P_1 = P_2 + P_{L2} + R\left(\frac{(P_2 + P_{L2})^2 + (Q_2 + Q_{L2})^2}{|V_2|^2}\right) \tag{1}$$

$$Q_1 = Q_2 + Q_{L2} + X\left(\frac{(P_2 + P_{L2})^2 + (Q_2 + Q_{L2})^2}{|V_2|^2}\right) \tag{2}$$

Then, the voltage magnitude of bus (r) can be determined in a forward direction as follows:

$$V_2^2 = V_1^2 - 2(P_1 R + Q_1 X) + \left(R^2 + X^2\right)\frac{(P_1{}^2 + Q_1{}^2)}{V_1^2} \tag{3}$$

The problem formulation can be presented as a multiobjective function as follows:

$$f_t = k_1 f_1 + k_2 f_2 + k_3 f_3 \tag{4}$$

where,

$$f_1 = \sum_{m=1}^{B}(P_{loss}(m)) \tag{5}$$

$$f_2 = \sum_{m=1}^{S}(VD(m)) \tag{6}$$

$$f_3 = \frac{1}{\sum_{m=1}^{S}(|VSI(m)|)} \tag{7}$$

$$|k_1| + |k_2| + |k_3| = 1 \tag{8}$$

where, $VD(m)$ represents the voltage deviation at bus (m), $VSI(m)$ represents the voltage stability index at bus (m), S is the total system buses, and B is the total system branches. K_1, K_2, and K_3 are weighting factors that are equal to 0.5, 0.25, and 0.25, respectively.

The inequality and equality constraints are determined as shown next [19,25,26].

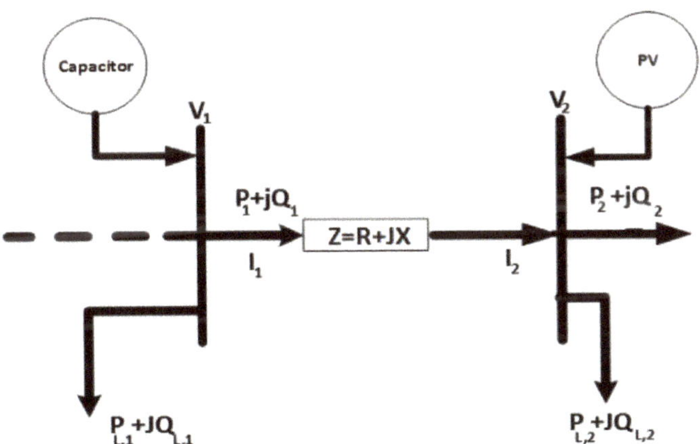

Figure 1. Representation of two buses in a distribution network.

2.1. Equality Constraints

The power flow balance equation that can be represented by (9) and (10), and the power flow equation that can be represented by (11) and (12), are the inequality constraints as shown next.

$$P_S + \sum_{m=1}^{M} P_{PV}(m) = \sum_{m=1}^{S} P_{L,m} + \sum_{m=1}^{B} P_{loss}(m) \tag{9}$$

$$Q_S + \sum_{m=1}^{N} Q_{Capacitor}(m) = \sum_{m=1}^{S} Q_{L,m} + \sum_{m=1}^{B} Q_{loss}(m) \tag{10}$$

$$P_1 = P_2 + P_{L2} + R\left(\frac{(P_2 + P_{L2})^2 + (Q_2 + Q_{L2})^2}{|V_2|^2}\right) \tag{11}$$

$$Q_1 = Q_2 + Q_{L2} + X\left(\frac{(P_2 + P_{L2})^2 + (Q_2 + Q_{L2})^2}{|V_2|^2}\right) \tag{12}$$

where, Q_S and P_S are the output reactive and real power from the grid, respectively, but $Q_{L,m}$ and $P_{L,m}$ are the reactive and real load demand at bus (m), respectively. M and N are the total number of PVs and capacitors in RDS, respectively. $P_{PV}(m)$ and $Q_{capacitor}(m)$ are the output power of PVs and capacitors at bus (m), respectively.

2.2. Inequality Constraints

The system operation constraints are the inequality constraints, which can be represented as follows:

2.2.1. System Voltage Constraints

The bus system voltage is operating between the minimum operating voltage (V_{Down}) and the maximum operating voltage (V_{Up}).

$$V_{down} \le V_m \le V_{up} \tag{13}$$

DER Sizing Limits:

$$\sum_{m=1}^{M} P_{PV}(m) \le \left(\sum_{m=1}^{S} P_{L,m} + \sum_{m=1}^{B} P_{loss}(m)\right) \tag{14}$$

$$P_{PV,\,n} \le P_{PV} \le P_{PV,a} \tag{15}$$

The PV output is operating between the minimum ($P_{PV,n}$) and maximum power ($P_{PV,a}$) of PVs in RDS.

Capacitor Size Limits:

$$\sum_{m=1}^{N} Q_{Capacitor}(m) \le \left(\sum_{m=1}^{S} Q_{L,\,m} + \sum_{m=1}^{B} Q_{loss}(m)\right) \tag{16}$$

$$Q_{Capacitor,\,n} \le Q_{Capacitor} \le Q_{Capacitor,a} \tag{17}$$

2.2.2. Line Capacity Limits

The branches current of the system is operating under operating constraints:

$$I_m \le I_{a,m} \ m = 1,\ 2,\ 3,\ \ldots,\ \text{NBr} \tag{18}$$

where, $I_{a,m}$ represents the high operating current in the branch (m).

3. TSA-SCA Algorithm

3.1. Tunicate Swarm Algorithm (TSA)

Tunicates are cylinder-shaped, with a gelatinous tunic that is closed at one end and open at the other. Tunicates shine with bioluminescence, generating a faint green-blue light, which can be viewed from several meters away. The size of tunicates is a few millimeters. In the ocean, tunicates absorb water to generate jet propulsion from their open ends using atrial siphons. Tunicates move in water by generating jet propulsion. The updating position of tunicates can be formulated as follow:

$$\vec{P_p}(x) = \begin{cases} \vec{FS} + \vec{A} \cdot \vec{PD} & ,if \ r_{rand} \ge 0.5 \\ \vec{FS} - \vec{A} \cdot \vec{PD} & ,if \ r_{rand} < 0.5 \end{cases} \tag{19}$$

$$\vec{M} = |P_{min} + c1 \cdot (P_{max} - P_{min})| \tag{20}$$

$$\vec{A} = \frac{\vec{G}}{\vec{M}} \tag{21}$$

$$\vec{G} = c2 + c3 - \vec{F} \tag{22}$$

$$\vec{F} = 2c1 \tag{23}$$

$$\vec{PD} = \left| \vec{FS} - r_{rand} \cdot \vec{P_p}(x) \right| \tag{24}$$

Tunicates move in a swarm in nature, which can be modeled by the following equation:

$$\vec{P_p}(x+1) = \frac{\vec{P_p}(x) + \vec{P_p}(x+1)}{2 + c1} \tag{25}$$

3.2. Sine-Cosine Algorithm (SCA)

The SCA is derived from the cosine and sine function to create an effective optimization algorithm. The effectiveness of the SCA is based on its exploitation and exploration phases. r_1 is used to balance the exploitation and exploration rates for the SCA over the course of iterations to obtain the global optimum solutions. The position of the SCA is updated as follows:

$$X_i^T + 1 = X_i^T + r_1 \cos(r_2) * \left| r_3 * P_i^T - X_i^T \right|, \; r_4 \geq 0.5 \tag{26}$$

$$X_i^T + 1 = X_i^T + r_1 \cos(r_2) * \left| r_3 * P_i^T - X_i^T \right|, \; r_4 < 0.5 \tag{27}$$

$$r_1 = 2 - \frac{2T}{T_{max}} \tag{28}$$

$$r_2 = 2 * \text{rand}() \tag{29}$$

$$r_3 = 2 * \text{rand}() \tag{30}$$

$$r_4 = \text{rand}() \tag{31}$$

Where, T_{max} and T are the maximum and current iteration, respectively, P_i^T is the targeted global optimal solution, and X_i^T represents the current iteration. r_1, r_2, r_3, and r_4 are random numbers.

3.3. Improved TSA-SCA Algorithm

The improved TSA-SCA is created by applying the updating position of the SCA to the updating position of the TSA to improve the exploration phase of the TSA. The rest of the pseudo code of the TSA remains the same, as shown in the following equations:

$$\vec{PD} = rand() * \sin(rand()) \left| \vec{FS} - r_{rand} \cdot \vec{P_p}(x) \right|, \; \text{rand}() < 0.5 \tag{32}$$

$$\vec{PD} = rand() * \cos(rand()) \left| \vec{FS} - r_{rand} \cdot \vec{P_p}(x) \right|, \; \text{rand}() < 0.5 \tag{33}$$

$$\vec{P_p}(x) = \begin{cases} \vec{FS} + \vec{A} \cdot \vec{PD} \;, \; if \; r_{rand} \geq 0.5 \\ \vec{FS} - \vec{A} \cdot \vec{PD} \;, \; if \; r_{rand} < 0.5 \end{cases} \tag{34}$$

$$\vec{P_p}(x+1) = \frac{\vec{P_p}(x) + \vec{P_p}(x+1)}{2 + c1} \tag{35}$$

The steps of the TSA-SCA to determine the optimal sizes and locations of DG in distribution networks are explained in the following steps:

1. Read the system data, maximum iteration (I), and number of search agents (S).
2. Produce the initial population of slime mold between the lower-(w) and upper (p)-controlled variables by Equation (36).

$$J(c,q) = rand(R(c,q) - y(c,q)) + y(c,q) \tag{36}$$

where, *rand* represents a random value between the values of 0 and 1. c and q are the number of tunicates and the problem dimension.

3. The produced population represents the tunicate position that can be formulated as follows:

$$S = \begin{bmatrix} S_{1,1} & S_{1,2} & S_{1,q-1} & S_{1,q} \\ S_{2,1} & \cdots & S_{2,q-1} & S_{2,q} \\ \vdots & \ddots & \vdots & \vdots \\ S_{N,1} & \cdots & S_{N,q-1} & S_{N,q} \end{bmatrix} \tag{37}$$

where, $S_{i,j}$ is the position of the tunicate.

4. Evaluate the fitness for all locations of tunicates, and obtain the superior position of tunicates and the superior objective function.
5. Evaluate the new position of each tunicate by Equation (35).
6. Return to step 4 until the final iteration is reached.
7. Obtain the best location of tunicates (sizes and positions of DG).

4. Testing and Evaluation

As mentioned in Section 1, the proposed optimization method is applied to the IEEE 69-node RDS to identify the optimal sizing and allocation of DG in such a system, considering the minimization of power losses. Such a system is tested by the proposed method when installing one, two, and three DGs to it. Also, the DG mode changes, i.e., DG is tested in three modes: while generating active power only (P-type mode), while generating reactive power only (Q-type mode), and while generating both active and reactive power.

The presented test system is a IEEE 69-node RDS that includes load demand of 3801.49 kW and 2694.6 kVAR [23]. This system consists of 69 buses and 68 branches with a base of 12.66 kV and 100 MVA. Figure 2 shows the system implementation by NEPLAN software.

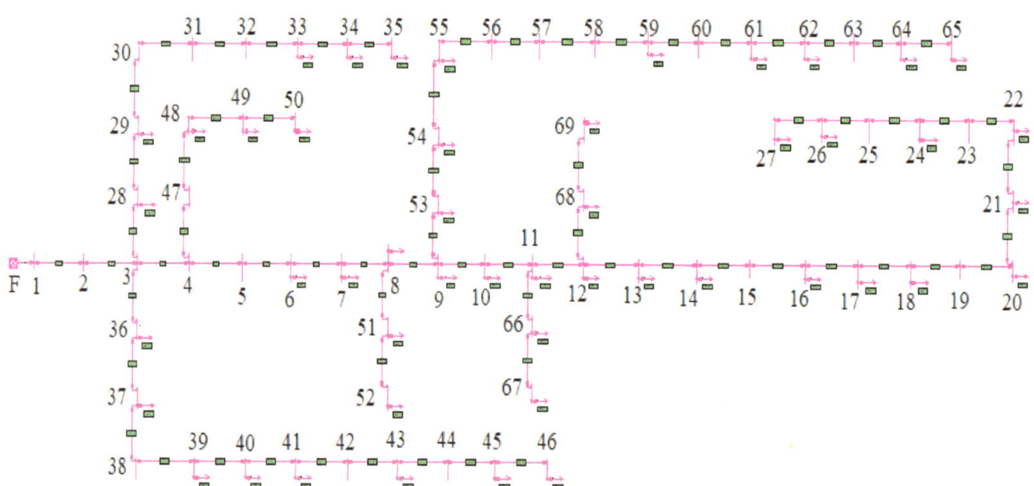

Figure 2. IEEE 69 RDS implementation by NEPLAN.

The analysis of the IEEE 69-node RDS system in its original case shows the active power losses of the system (P_{Losses} = 225 kW), and the least voltage is in node 65, where the voltage of that node is V_{65} = 0.9092 pu. Figure 3 shows the loading ability of the system by PV curves, where PV curves are applied on buses 10, 27, 46, and 65.

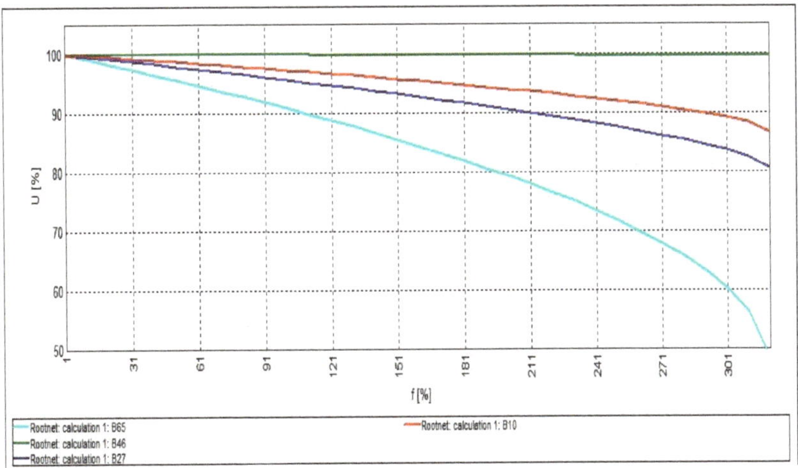

Figure 3. PV curves of the IEEE 69-node RDS system in its original case.

The PV curves show that the voltage of node 65—the weakest node in the RDS—starts to fall below 90% of nominal voltage when loads exceed 110%, and the system collapses when loads exceed 320%. Now the hybrid TSA-SCA method can start to be used to optimize the system by identifying the optimal allocation and sizing of DG installed regarding active power losses. After the optimization process, the system is analyzed to show the impact of the optimization process on the performance of the system. The optimization process is achieved using MATLAB R2019b on a personal computer with 2 GB RAM and Intel(R) Pentium(R) CPU B950, 2.1 GHZ, while the analyses are achieved using NEPLAN software.

4.1. System Optimization with Active Power-Generating DG (P-Type Mode)

In this case, the DGs installed are P-type and are configured for system optimization regarding the objective function as it is shown in Table 1.

Table 1. Configuration of P-type DGs installed on the IEEE-69 RDS.

Number of DGs	1	2	3
Location/Size (kW)	61/1873.32	61/1781.2 17/530.5	61/1718.8 17/380.5 11/525
Power Losses (kW)	83.2224	71.6745	69. 4266
Power Losses Reduction (%)	63.012	68.145	69.144

According to the settings of DGs shown in Table 1, the load flow process and PV curves were carried out. Starting with the system with one DG installation, results show power loss (Ploss = 83.2224 kW), and the total power loss reduction = 63.012%. The least voltage is found in node 27, where V_{27} = 0.9683 pu. Figure 4 shows the system's PV curves with one P-type DG installation.

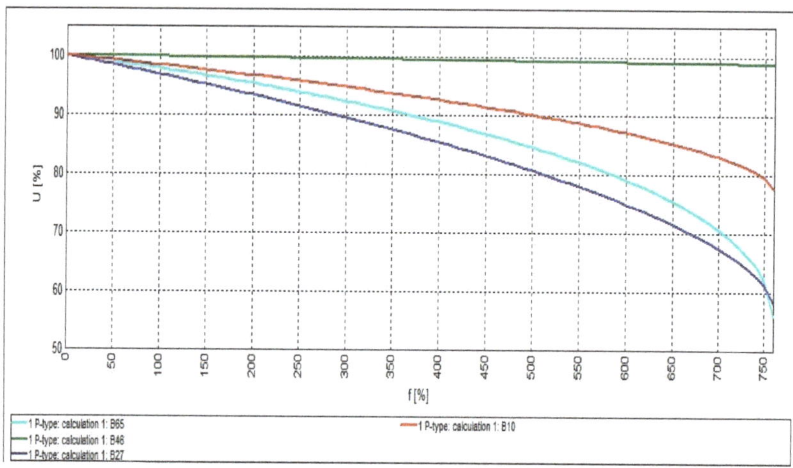

Figure 4. PV curves of the IEEE 69-node RDS system with 1 P-type DG installation.

Due to the new DG's location, the PV curve of node 65 improved, as shown in Figure 4. Figure 4 shows that the voltage of node 27—the system's weakest node in that case—starts to fall below 90% of nominal voltage when loads exceed 290%, and the system collapses when loads exceed 750%.

When installing two P-type DGs to the system according to the settings in Table 1, the active power loss of the system is P_{Loss} = 71.6745 kW, and the total active power loss reduction is 68.145%. The least voltage is found in node 65, where V_{65} = 0.9789 pu. Figure 5 shows the system's PV curves with two P-type DG installations.

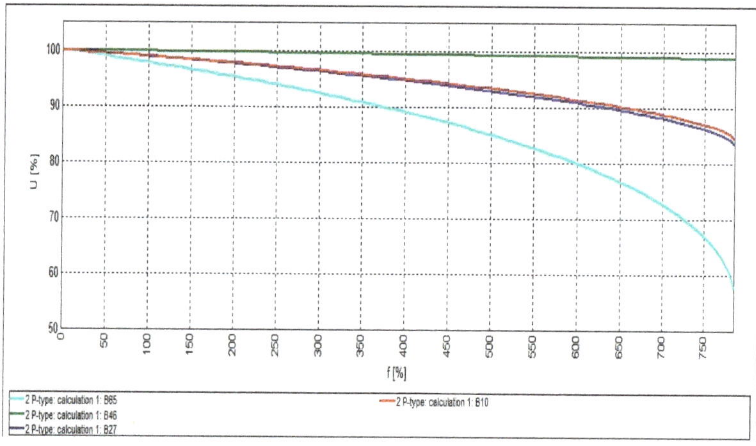

Figure 5. PV curves of the IEEE 69-node RDS system with 2 P-type DG installations.

The presence of the two DGs in the estimated locations improves the PV curves even more. Figure 5 shows that the voltage of node 65—the weakest node in that case—starts to fall below 90% of nominal voltage when loads exceed 380%, and the system collapses when loads exceed 750%.

When installing three P-type DGs to the system according to the settings of Table 1, the active power loss of the system is P_{Loss} = 69.4266 kW, and the total active power loss reduction is 69.144%. The least voltage is found in node 65, where V_{65} = 0.979 pu. Figure 6 presents the system's PV curves with three P-type DG installations.

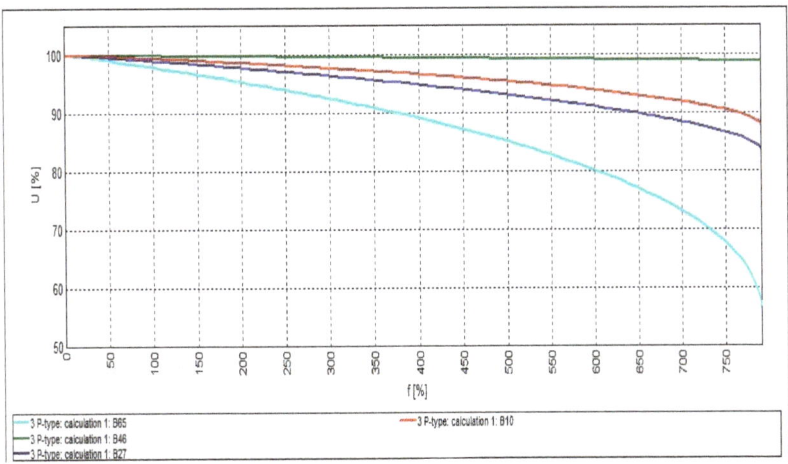

Figure 6. PV curves of the IEEE 69-node RDS system with 3 P-type DG installations.

Figure 6 shows that there is almost no drastic effect on the system's PV curves when installing 3 P-type DGs. The voltage of node 65—the system's weakest node in that case—starts to fall below 90% of nominal voltage when loads exceed 380%, and the system collapses when loads exceed 750%. It is also observed that the presence of the third DG resulted in an improvement on node 10's PV curve.

4.2. System Optimization with Reactive Power Generating DGs (Q-Type Mode)

In this case, the DGs installed are Q-type DGs and are configured for system optimization regarding the objective function as it is shown in Table 2.

Table 2. Configuration of Q-type DGs installed in the IEEE-69 RDS.

Number of DGs	1	2	3
Location/Size (kVAr)	61/1330	61/1276 17/361	61/1233 17/253 11/391
Power Losses (kW)	152.041	146.441	145.129
Power Losses Reduction (%)	32.426	34.915	35.498

Following the settings of DGs shown in Table 2, the load flow process and PV curves were carried out. When installing one Q-type DG, the power loss of the system is P_{Loss} = 152.041 kW, and the total power loss reduction = 32.426%. The least voltage is found in node 65, where V_{27} = 0.9307 pu. Figure 7 presents the system's PV curves with one Q-type DG installation.

In the presence of one Q-type DG, the system's PV curves improved. Figure 7 shows that the voltage of node 65—the system's weakest node in that case—starts to fall below 90% of nominal voltage when loads exceeds 140%, and the system collapses when loads exceeds 400%.

When installing two Q-type DGs to the system according to the settings in Table 2, the active power loss of the system is P_{Loss} = 146.441 kW, and the total active power loss reduction is 34.915%. The least voltage is found in node 65, where V_{65} = 0.9311 pu. Figure 8 presents the system's PV curves with two Q-type DG installations.

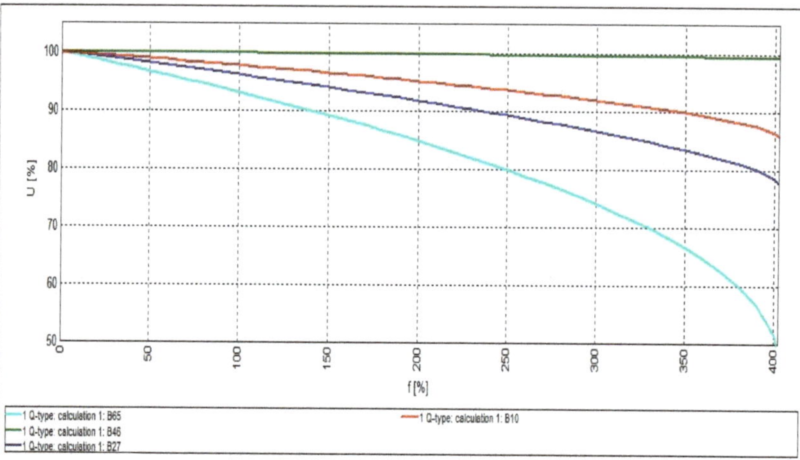

Figure 7. PV curves of the IEEE 69-node RDS system with 1 Q-type DG installation.

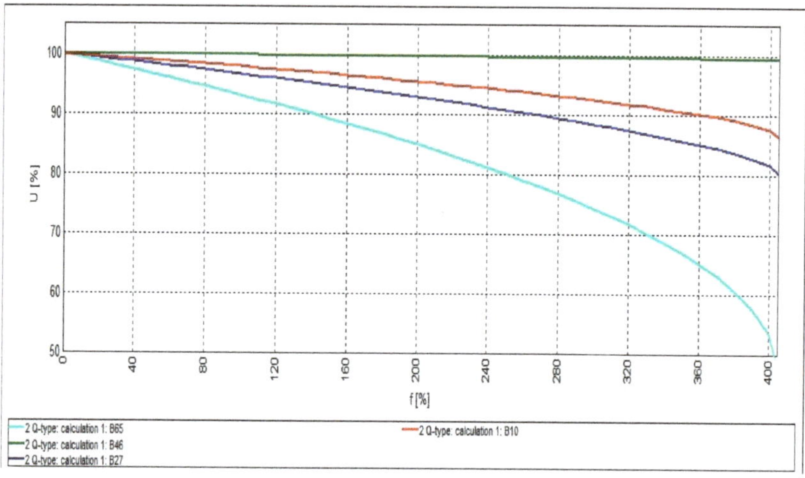

Figure 8. PV curves of the IEEE 69-node RDS system with 2 Q-type DG installations.

Figure 8 shows that there is almost no drastic effect on the system's PV curves when installing 2 Q-type DGs. The voltage of node 65—the system's weakest node in that case—starts to fall below 90% of nominal voltage when loads exceed 140%, and the system collapses when loads exceed 400%.

When installing three Q-type DGs to the system according to the settings of Table 2, the active power loss of the system is P_{Loss} = 145.129 kW, and the total active power loss reduction is 35.498%. The least voltage is found in node 65, where V_{65} = 0.9314 pu. Figure 9 presents the system's PV curves with three Q-type DG installations.

Again, Figure 9 shows that there is almost no drastic effect on the system's PV curves when installing 3 Q-type DGs. The voltage of node 65—the system's weakest node in that case—starts to fall below 90% of nominal voltage when loads exceed 140%, and the system collapses when loads exceed 400%.

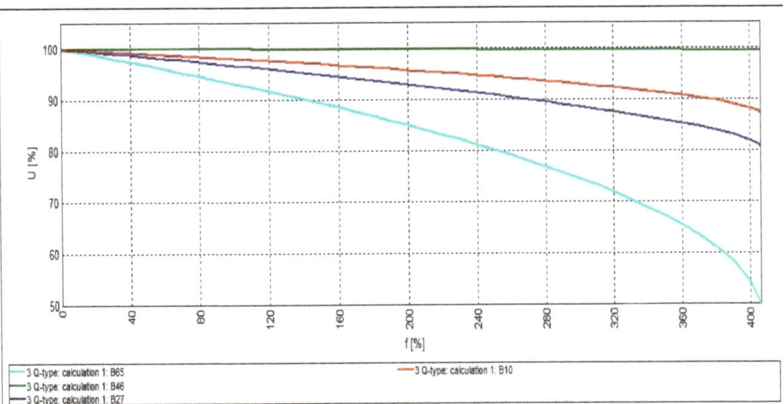

Figure 9. PV curves of the IEEE 69-node RDS system with 3 Q-type DG installations.

4.3. System Optimization with Active and Reactive Power-Generating DG (PQ-Type Mode)

In this case, the DGs installed are PQ-type DGs and are configured for system optimization regarding the objective function as it is shown in Table 3.

Table 3. Configuration of PQ-type DG installed on the IEEE-69 RDS.

Number of DGs	1	2	3
Location/Size (kVAr)/P.F	61/1828.44/0,8149	61/1735/0.8138 17/523.24/0.829	61/1673.2/0.8136 17/377.86/0.8312 11/497.33/0.8155
Power Losses (kW)	23.169	7.2013	4.2665
Power Losses Reduction (%)	89.702	96.799	98.104

Following the settings of DGs shown in Table 3, the load flow process and PV curves were carried out. When installing one PQ-type DG, the power losses, P_{Loss} = 23.169 kW, and the total power loss reduction = 89.702%. The least voltage is found in node 27, where V_{27} = 0.9725 pu. Figure 10 presents the system's PV curves with one PQ-type DG installation.

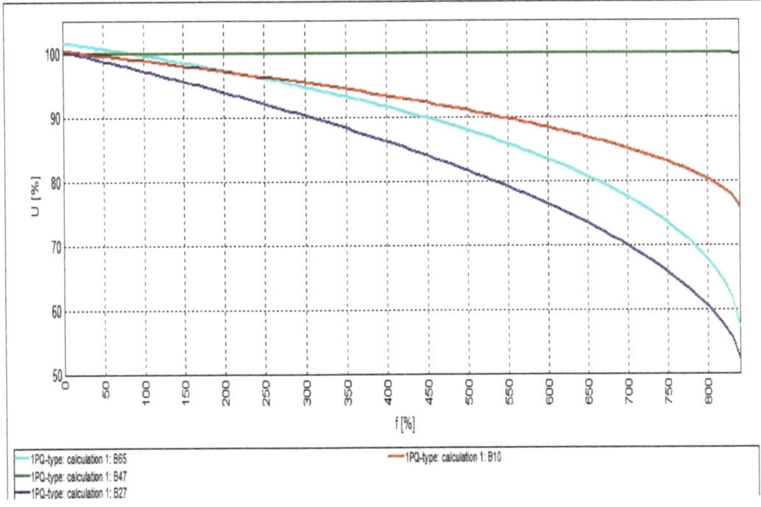

Figure 10. PV curves of the IEEE 69-node RDS system with 1 PQ-type DG installation.

Due to the presence of the powerful PQ-type DG, the system's PV curves acquired superior improvement, as the DG provides both active and reactive power to the RDS. Figure 10 shows that the voltage of node 27—the system's weakest node in that case—starts to fall below 90% of nominal voltage when loads exceed 310%, and the system collapses when loads exceed 850%.

When installing two PQ-type DGs to the system according to the settings in Table 3, the active power loss of the system is P_{Loss} = 7.2013 kW, and the total active power loss reduction is 96.799%. The least voltage is found in node 69, where V_{69} = 0.9943 pu. Figure 11 presents the system's PV curves with two PQ-type DG installations.

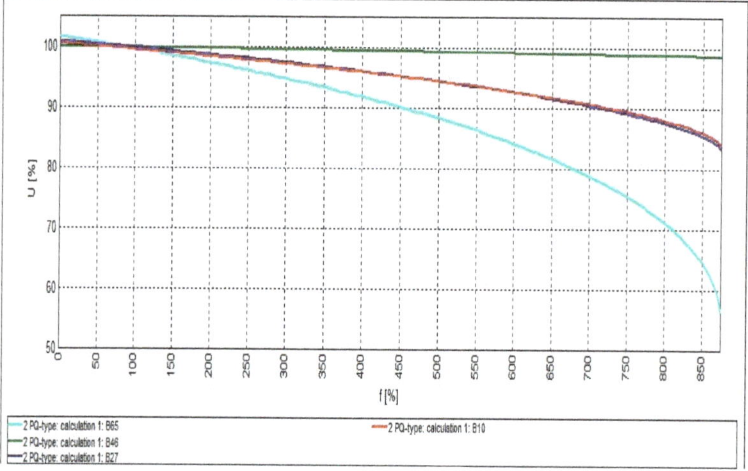

Figure 11. PV curves of the IEEE 69-node RDS system with 2 PQ-type DG installations.

Due to the presence of the two PQ-type DGs, the system's PV curves were improved even more. Figure 11 shows that the voltage of node 65—the system's weakest node in that case—start to fall below 90% of nominal voltage when loads exceed 450%, and the system collapses when loads exceed 850%.

When installing three PQ-type DGs to the system according to the settings of Table 3, the active power loss of the system is P_{Loss} = 4.2665 kW, and the total active power loss reduction is 98.104%. The least voltage is found in node 65, where V_{65} = 0.997 pu. Figure 12 presents the system's PV curves with three PQ-type DG installations.

Figure 12 shows that a slight increase in the system's load capacity happens when installing 3 PQ-type DGs in the system. The voltage of node 65—the system's weakest node in that case—starts to fall below 90% of nominal voltage when loads exceed 460%, and the system collapses when loads exceed 850%. Even more, the PV curve of node 10 is slightly improved more than in the case of two PQ-mode DGs.

Table 4 shows the comprehensive results between TSA-SCA techniques and several other techniques.

Table 4. Comparison results of the improved TSA-SCA algorithm and other algorithms in the IEEE 69-bus RDS.

Item	TSA-SCA	MFO [8]	Hybrid [27]	WOA [28]	SCA [29]	PSO [30]	PVSC [31]
1P-type	83.2224	83.224	83.372	-	-	-	-
2P-type	71.6745	71.679	71.82	-	-	-	-
3P-type	69.4266	-	69.52	-	-	-	-
1Q-type	152.041	-	-	152.064	-	-	-
2Q-type	146.441	-	-	-	147.762	-	-
3Q-type	145.129	-	-	-	-	-	-
1PQ-type	23.169	-	-	-	-	25.9	-
2PQ-type	7.201	-	-	-	-	-	-
3PQ-type	4.253	-	-	-	-	-	9.63

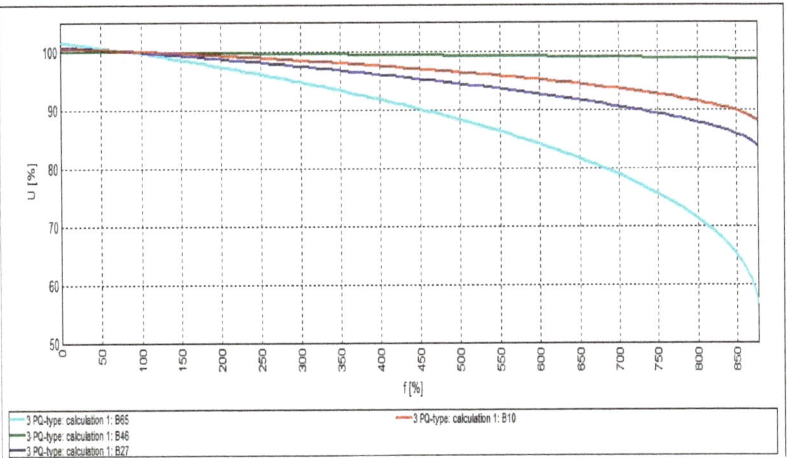

Figure 12. PV curves of the IEEE 69-node RDS system with 3 PQ-type DG installations.

From Table 4, the improved TSA-SCA has efficient characteristics to obtain the best results when compared to other efficient algorithms.

5. Conclusions

This paper introduces a new hybrid approach, a mixture between the tunicate swarm algorithm (TSA) and the sine-cosine algorithm (SCA) optimization technique. The new TSA/SCA optimization technique was tested on an IEEE 69-node RDS system, where the fitness is decreasing the active power loss through identifying the optimal sizing and allocation of DGs installed on the system. The optimization process took place in three modes: with active power DGs (P-mode DGs), reactive power DGs (Q-mode DGs), and with both active and reactive power DGs (PQ-mode DGs). The performance of the optimized system was evaluated by NEPLAN software to show the impact of the optimization process on the system. The analyses shows that the objective function was successfully achieved by the proposed technique, the active power loss was obviously minimized and the load demand of the system was greatly increased so that it can withstand more loads. The results proved that integration of multiple DGs gives better results than the integration of a single DG in a distribution network. Installing PQ-type optimization gives better results than the integration of P-type or Q-type.

Author Contributions: Conceptualization, A.A., H.A.-M. and S.K.; Data curation, A.A.I. and F.J.; Formal analysis, S.K.; Resources, A.A. and H.A.-M.; Methodology, A.A.I. and F.J.; Software, A.A., H.A.-M. and S.K.; Supervision, A.A.I. and F.J.; Validation, A.A. and S.K.; Visualization, H.A.-M., A.A.I. and F.J.; Writing—original draft, A.A., H.A.-M. and S.K.; Writing—review & edit-ing, A.A.I. and F.J. All authors together organized and refined the manuscript in the present form. All authors have approved the final version of the submitted paper. All authors have read and agreed to the published version of the manuscript.

Funding: This research was funded by NSFC (China)-ASRT (Egypt) Joint Research Fund grant number 51861145406.

Institutional Review Board Statement: Not applicable.

Informed Consent Statement: Not applicable.

Data Availability Statement: Not applicable.

Acknowledgments: The authors gratefully acknowledge the contribution of the NSFC (China)-ASRT (Egypt) Joint Research Fund, Project No. 51861145406 for providing partial research funding to the work reported in this research.

Conflicts of Interest: The authors declare no conflict of interest.

References

1. Abdmouleh, Z.; Gastli, A.; Ben-Brahim, L.; Haouari, M.; Al-Emadi, N.A. Review of optimization techniques applied for the integration of distributed generation from renewable energy sources. *Renew. Energy* **2017**, *113*, 266–280. [CrossRef]
2. Mekhilef, S.; Saidur, R.; Kamalisarvestani, M. Effect of dust, humidity and air velocity on efficiency of photovoltaic cells. *Renew. Sustain. Energy Rev.* **2012**, *16*, 2920–2925. [CrossRef]
3. Rizzi, F.; van Eck, N.J.; Frey, M. The production of scientific knowledge on renewable energies: Worldwide trends, dynamics and challenges and implications for management. *Renew. Energy* **2014**, *62*, 657–671. [CrossRef]
4. Paska, J.; Biczel, P.; Kłos, M. Hybrid power systems–An effective way of utilising primary energy sources. *Renew. Energy* **2009**, *34*, 2414–2421. [CrossRef]
5. Raheem, A.; Abbasi, S.A.; Memon, A.; Samo, S.R.; Taufiq-Yap, Y.H.; Danquah, M.K.; Harun, R. Renewable energy deployment to combat energy crisis in Pakistan. *Energy Sustain. Soc.* **2016**, *6*, 1–13. [CrossRef]
6. Ashfaq, A.; Ianakiev, A. Features of fully integrated renewable energy atlas for Pakistan; wind, solar and cooling. *Renew. Sustain. Energy Rev.* **2018**, *97*, 14–27. [CrossRef]
7. Zaharim, A.; Sopian, K. Prospects of life cycle assessment of renewable energy from solar photovoltaic technologies: A review. *Renew. Sustain. Energy Rev.* **2018**, *96*, 11–28.
8. Abdel-mawgoud, H.; Kamel, S.; Ebeed, M.; Aly, M.M. An efficient hybrid approach for optimal allocation of DG in radial distribution networks. In Proceedings of the 2018 International Conference on Innovative Trends in Computer Engineering (ITCE), Aswan, Egypt, 19–21 February 2018; pp. 311–316.
9. Engelen, P.J.; Kool, C.; Li, Y. A barrier options approach to modeling project failure: The case of hydrogen fuel infrastructure. *Resour. Energy Econ.* **2016**, *43*, 33–56. [CrossRef]
10. Ranieri, L.; Mossa, G.; Pellegrino, R.; Digiesi, S. Energy recovery from the organic fraction of municipal solid waste: A real options-based facility assessment. *Sustainability* **2018**, *10*, 368. [CrossRef]
11. Di Corato, L.; Moretto, M. Investing in biogas: Timing, technological choice and the value of flexibility from input mix. *Energy Econ.* **2011**, *33*, 1186–1193. [CrossRef]
12. Huisman, K.J.; Kort, P.M. Strategic technology adoption taking into account future technological improvements: A real options approach. *Eur. J. Oper. Res.* **2004**, *159*, 705–728. [CrossRef]
13. Kaur, S.; Awasthi, L.K.; Sangal, A.L.; Dhiman, G. Tunicate swarm algorithm: A new bio-inspired based metaheuristic paradigm for global optimization. *Eng. Appl. Artif. Intell.* **2020**, *90*, 103541. [CrossRef]
14. Mirjalili, S. SCA: A sine cosine algorithm for solving optimization problems. *Knowl. Based Syst.* **2016**, *96*, 120–133. [CrossRef]
15. Reddy, P.D.P.; Reddy, V.V.; Manohar, T.G. Whale optimization algorithm for optimal sizing of renewable resources for loss reduction in distribution systems. *Renew. Wind Water Sol.* **2017**, *4*, 1–13. [CrossRef]
16. Thangaraj, Y.; Kuppan, R. Multi-objective simultaneous placement of DG and DSTATCOM using novel lightning search algorithm. *J. Appl. Res. Technol.* **2017**, *15*, 477–491. [CrossRef]
17. Syed, M.S.; Injeti, S.K. Simultaneous optimal placement of DGs and fixed capacitor banks in radial distribution systems using BSA optimization. *Int. J. Comput. Appl.* **2014**, *108*.
18. Lalitha, M.P.; Babu, P.S.; Adivesh, B. Optimal distributed generation and capacitor placement for loss minimization and voltage profile improvement using Symbiotic Organisms Search Algorithm. *Int. J. Electr. Eng.* **2016**, *9*, 249–261.
19. Barati, H.; Shahsavari, M. Simultaneous Optimal placement and sizing of distributed generation resources and shunt capacitors in radial distribution systems using Crow Search Algorithm. *Int. J. Ind. Electron. Control Optim.* **2018**, *1*, 27–40.
20. Krueasuk, W.; Ongsakul, W. Optimal placement of distributed generation using particle swarm optimization. In Proceedings of the Australian Universities Power Engineering Conference, Melbourne, Australia, 10–13 December 2006; pp. 10–13.
21. El-Fergany, A. Optimal allocation of multi-type distributed generators using backtracking search optimization algorithm. *Int. J. Electr. Power Energy Syst.* **2015**, *64*, 1197–1205. [CrossRef]
22. Kumar Injeti, S.; Shareef, S.M.; Kumar, T.V. Optimal allocation of DGs and capacitor banks in radial distribution systems. *Distrib. Gener. Altern. Energy J.* **2018**, *33*, 6–34. [CrossRef]
23. Abdelaziz, A.Y.; Ali, E.S.; Abd Elazim, S.M. Flower pollination algorithm and loss sensitivity factors for optimal sizing and placement of capacitors in radial distribution systems. *Int. J. Electr. Power Energy Syst.* **2016**, *78*, 207–214. [CrossRef]
24. Eminoglu, U.; Hocaoglu, M.H. Distribution systems forward/backward sweep-based power flow algorithms: A review and comparison study. *Electr. Power Compon. Syst.* **2008**, *37*, 91–110. [CrossRef]
25. Abdel-mawgoud, H.; Kamel, S.; Ebeed, M.; Youssef, A.R. Optimal allocation of renewable dg sources in distribution networks considering load growth. In Proceedings of the 2017 Nineteenth International Middle East Power Systems Conference (MEPCON), Cairo, Egypt, 19–21 December 2017; pp. 1236–1241.
26. Ali, E.S.; Abd Elazim, S.M.; Abdelaziz, A.Y. Ant lion optimization algorithm for renewable distributed generations. *Energy* **2016**, *116*, 445–458. [CrossRef]
27. Kansal, S.; Kumar, V.; Tyagi, B. Hybrid approach for optimal placement of multiple DGs of multiple types in distribution networks. *Int. J. Electr. Power Energy Syst.* **2016**, *75*, 226–235. [CrossRef]

28. Reddy, V.V.C. Optimal renewable resources placement in distribution networks by combined power loss index and whale optimization algorithms. *J. Electr. Syst. Inf. Technol.* **2018**, *5*, 175–191.
29. Biswal, S.R.; Shankar, G. Optimal Sizing and Allocation of Capacitors in Radial Distribution System using Sine Cosine Algorithm. In Proceedings of the 2018 IEEE International Conference on Power Electronics, Drives and Energy Systems (PEDES), Chennai, India, 18–21 December 2018; pp. 1–4.
30. Aman, M.M.; Jasmon, G.B.; Solangi, K.H.; Bakar, A.H.A.; Mokhlis, H. Optimum simultaneous DG and capacitor placement on the basis of minimization of power losses. *Int. J. Comput. Electr. Eng.* **2013**, *5*, 516. [CrossRef]
31. Nawaz, S.A.R.F.A.R.A.Z.; Bansal, A.K.; Sharma, M.P. Optimal Allocation of Multiple DGs and Capacitor Banks in Distribution Network. *Eur. J. Sci. Res.* **2016**, *142*, 153–162.

clean technologies

Article

Renewable Energy at Home: A Look into Purchasing a Wind Turbine for Home Use—The Cost of Blindly Relying on One Tool in Decision Making

Sheridan Ribbing [1] **and George Xydis** [1,2,*]

[1] Energy Policy and Climate Program, Krieger School of Arts & Sciences, Johns Hopkins University, Baltimore, MD 21218, USA; sheridanribbing@yahoo.com

[2] Department of Business Development and Technology, Aarhus University, Birk Centerpark 15, 7400 Herning, Denmark

* Correspondence: gxydis@gmail.com or gxydis@btech.au.dk; Tel.: +45-93508006

Abstract: Small-scale wind turbines simulations are not as accurate when it comes to costs as compared to the large-scale wind turbines, where costs are more or less standard. In this paper, an analysis was done on a decision for a wind turbine investment in Bellingham, Whatcom County, Washington. It was revealed that a decision taken based only on a software tool could be destructive for the sustainability of a project, since not taking into account specific taxation, net metering, installation, maintenance costs, etc., beyond the optimization that the tool offers, can hide the truth.

Keywords: green communities; energy independence; HOMER; wind turbines

Citation: Ribbing, S.; Xydis, G. Renewable Energy at Home: A Look into Purchasing a Wind Turbine for Home Use—The Cost of Blindly Relying on One Tool in Decision Making. *Clean Technol.* **2021**, *3*, 299–310. https://doi.org/10.3390/cleantechnol3020017

Academic Editor: Patricia Luis

Received: 26 January 2021
Accepted: 28 February 2021
Published: 1 April 2021

Publisher's Note: MDPI stays neutral with regard to jurisdictional claims in published maps and institutional affiliations.

1. Introduction

1.1. Current Research Framework

Wind turbines have been at the forefront of renewable energy technology. Many Americans have noticed this development in pictures of Europe with tall white wind turbines scattered over green rolling hills. Many have seen the news from Texas developing large wind farms over miles of prairie lands, as well as of Block Island, Rhode Island, where the nation's first offshore wind turbine was recently installed [1]. The future of renewable energy in the United States continues to expand to residential backyards. After decades of wind turbine research and development, many European countries, such as Belgium and Denmark, lead the market with small wind turbines for private or community use— especially since Denmark's amazing turbine development began with community-bought wind turbines [2,3]

Wind turbines have recently become much cheaper, smaller, more efficient, and easier to transport and assemble [4–6]. This new technology allows families to purchase wind turbines for their homes and connect to the grid to be able to sell extra electricity back to the utility company or share with their neighbors. The Peer-to-Peer (P2P) approach from a computing application scheme has been made possible to be applied in other fields, such as in the renewable energy sector.

Generation Y, also referred to as Millennials, grew up during the birth of the internet. While this generation experienced a childhood similar to their parents—playing outside until the streetlights came on—the coming-of-age period of this generation occurred while the internet was being developed to function from people's hands, no longer on dial up, but on wireless cellphone computers. This was the time that humans began socializing in chat rooms, work began using emails, and you could search the internet for endless knowledge. The internet provided an amazing change in the way lay people could access information about the world around them. This generation would prove most concerned for the environment, as they absorbed much global information growing up [7]. Today, this generation have families and have bought or are looking to buy houses, and many are also looking to lower their carbon footprint—often not following the most efficient path. Electrification of their consumption is a top priority, including EV ownership [8], heat pumps, and other residential appliances [9]. Thus far, whenever energy surplus was generated via, e.g., a fireplace, via a fuel-based boiler etc., it was simply lost. Now, with P2P technology and the liberalization of the electricity markets (and how modern grids operate), it can be offered to the neighbor at a competitive price [10]. When the local utility company does not sell energy from renewable sources, there are options for families to do so themselves [11]. Over the last few years, in the Western world, the global Not-In-My-Back-Yard (NIMBY) approach has moved from there to Yes-In-My-Back-Yard (YIMBY). More and more are looking to be prosumers and—if in a warm climate area—are trying to invest in solar (or wind) energy resources and storage as much as possible, aiming at having a zero-electricity bill (considering—ultimately falsely—that they are at times grid-independent of the local utility) [12].

The renewable energy market offers a variety of wind turbines, solar panels, and biofuel options. Wind turbines are one of the older technologies that have undergone recent decades of research and development [13] and "is the fastest growing source of energy in the world—efficient, cost effective, and nonpolluting," according to [14], which makes it an ideal option for consumers, especially when paired with solar panels. Installing a wind turbine and/or solar panel requires research into the amount of energy used by the household, the laws of the local area and/or homeowners association, and the consumer's budget.

Although there is a large number of articles published focused on remote or rural areas, mostly in African countries [15–18], a significant amount of literature—though not extensive—is devoted to renewable energy at home, with a focus onto purchasing a wind turbine for home use. Oliver and Groulx [19] presented a homeowner-centric approach of a hybrid renewable energy system, which included a wind turbine, which proved what was already known for wind turbines: that they are clear economies of scale. Ugur et al. [20] moved on to a financial analysis for small wind turbines for home use in Turkey. Based on their results and the wind resource analysis in Konstantinoupoli, they have identified where the most profitable areas in the city for small wind installations are. Rodriguez-Hernandez et al. [21] did another economic feasibility study for small wind turbines in the Valley of Mexico metropolitan area, based on three years of data, 28 wind turbine models, and 18 locations. Hemmati [22] published a techno-economic analysis of a home system, which included a small-scale wind turbine and a storage subsystem. Mixed integer linear programming was used, and it was proven that the lowest planning costs were for a 20-kW wind turbine.

On the other hand, Canale et al. [23] were not focused on the economic analysis. Instead, they focused on an innovative blade technical analysis and their application on small-scale wind turbines. Numerical and experimental results were evaluated based on the Blade Element Momentum (BEM) theory in small wind turbines, which is not usually the case. Such experimental set-ups are usually met in large scale testing facilities. Another technical analysis was done for a 5-kW wind turbine system for a home with the inclusion of batteries [24]. A net-zero energy home was studied by Rasouli and Hemmati [25], by using mixed integer nonlinear programming (MINLP) and solved using the particle swarm optimization (PSO) technique. It was proven that any net zero energy home is heavily dependent on the wind turbine, solar sizing, battery sizing, and hydrogen (or in some cases electric) vehicles.

1.2. Renewable Energy at Home

What has not been studied adequately over the last years—and definitely not after the renewable energy's progress in the USA—is if people currently have a more positive attitude with regards to having a wind turbine in their back yard compared to the NIMBY approach, which has clearly lasted for a long time worldwide. It should be pointed out, however, that over the last year, the coronavirus pandemic derailed renewable energy's overall progress in several countries. The public deficit has increased, and the GDP, although it has experienced small ups and downs throughout the past decade (mainly ups, up to 5% compared to the previous calendar quarter every time), over the last two quarters in 2020 plummeted by −5.0% (Q1) and −32.9% (Q2) in the US [26] (Figure 1). Therefore, liquidity and available funds for investing in renewable energy sources for home use shrunk. Thus, the necessity to lower the initial capital investment of all renewables, mainly of small wind turbines, in order to achieve substantial growth in the small wind sector in addition to the large wind turbine sector is crucial. Furthermore, as we approach the American elections, budgetary-wise, liquidity for investments will become tougher to find and funds will be limited.

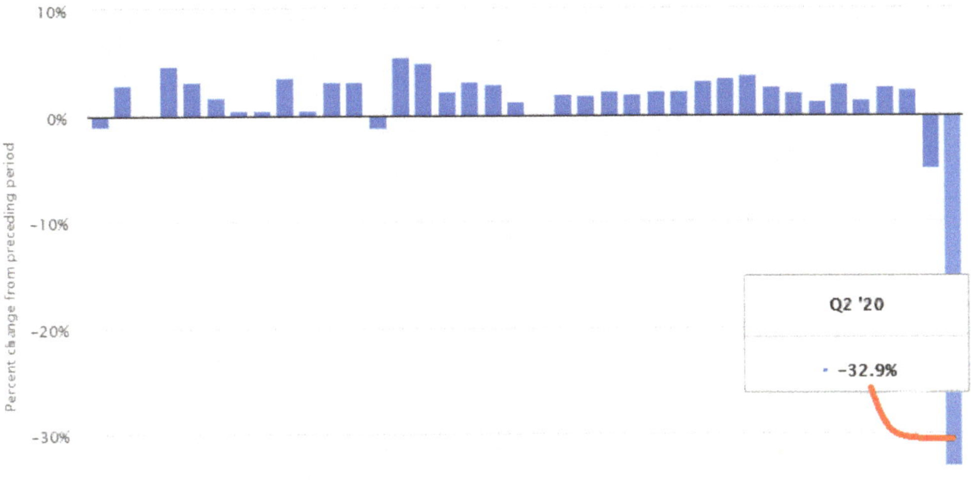

Figure 1. Quarterly growth of the real GDP in the United States from 2011 to 2020.

Beyond that, the insufficient infrastructure for small-scale investments, such for as small wind turbines, was the barrier for developing such business activities. However, Electric Vehicles (EVs), along with the P2P infrastructure, has led citizens to start thinking in another way. It is not only their will to produce their own wind energy and have a near-zero (or even negative) electricity bill towards energy independence, it is also a prestige-related attitude, the increase in social status, and the social acceptance related to the purchase of an EV and the expectance of its purchase to soon be a good value for money [27]. Moreover, such investments could follow the EV ownership approach. A "create-your-own-electricity" approach could be another label of social status. Such investments are linked more to summer houses than to houses in cities and densely urban areas, where the electricity demand is highly increased due to the increased flow of tourists. In fact, for some very hot days in summer when the grid is stressed, in order to fulfill the cooling demands, small wind turbines could contribute significantly to electricity generation. In those areas, there is a narrow security margin of electricity supply and a high risk of the system's breakdown in case of malfunctions/power cuts in periods with high load demand. Additionally, because of the different demand profile with the winter in the summer houses, the large daily and seasonal electricity load demand fluctuations, the summer peak can be many times greater than the lowest electricity demand in winter, and the daily fluctuation could be ±50–60% of the average value [28]. Approaching the above-mentioned issues from the perspective of microgrids and hybrid systems, many solutions can be found on several levels, even leading to a holistic energy climate water nexus [29,30]. Concerning financial issues, this will lead to energy savings and a reduction in operational costs. From another point of view, independent microgrids, which include wind turbines, should be able to operate as autonomous power networks in case of failure and forced isolation, providing uninterruptible power to crucial loads at least. In addition, microgrids with small wind turbines should be able to meet the peak loads and improve the power quality and stability of the grid-connection. To this end, the appropriate energy storage method (battery system) will play an important role.

Above all, ultimately, the goal is to create green communities, a large step towards energy independence and energy democracy. By introducing, e.g., small wind turbines via the P2P approach, the current grid-based electricity is changing. In such communities, the power comes from solar panels (on every roof) or small wind turbines, and is channeled to the same busbar, which is linked to energy storage, community heating, EV charging, etc. Energy is, therefore, consumed in the vicinity where it is produced, and the end-users do not need to worry about transmission losses.

2. Materials and Methods

The importance of software tools in decision making is high and is meaningful for all builders, architects, and decision makers. Software tools are helping them to allocate resources, make decisions for investments, decide on the companies' philosophy, etc. This is rather profound in the renewable energy sector. The work in this paper was supported by the HOMER tool (Figure 2). The HOMER Grid is an excellent software tool to help one discover the resource availability in the specific location of their choice. The program was designed by a retired Senior Economist at the U.S. Department of Energy's National Renewable Energy Laboratory [31]. The program offers not only resource information, but also options for specific models of wind turbines, solar panels, generators, and information from the utility company, alongside the associated costs and profits. Once a consumer determines their future address, the energy use of the household will be determined, and the HOMER Grid will find which systems would work best, after which it will find the local laws and apply for the building permit, and then contact the retailer to find the installation costs; finally, it will decide whether or not to invest.

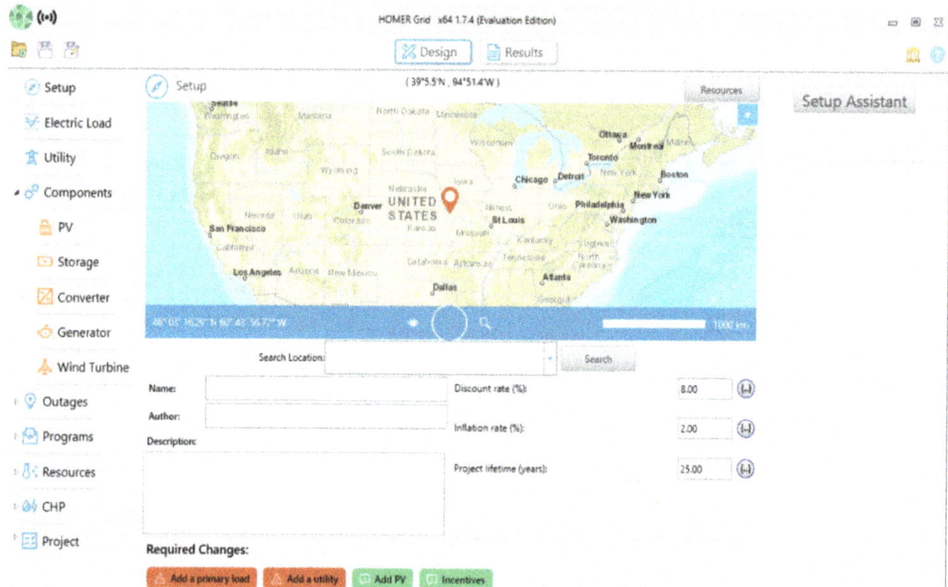

Figure 2. Screenshot from HOMER Grid showing the start screen when opening the application (2020).

This specific proposal is for a XANT M-21 100 kW to be installed in Bellingham, Whatcom County, Washington. This location was chosen as one option for a future home and a small farm for rescued animals. This turbine was introduced in the HOMER Grid program, which is the main component used in this proposal. This proposal and the steps taken to reach a conclusion may be helpful for any person curious about renewable energy for their home, or for communities looking into purchasing a wind turbine to share, as well as the types of people looking to live off the grid.

The wind turbine proposal will describe the steps taken to find the best results for the wind turbine purchase in this specific location in Bellingham, Washington, and an actual residence for sale on Zillow (Figure 3). Following the proposal, the reader will find a thorough analysis and explanations for the costs and cash flow. The analysis describes the few overlooked obstacles in the HOMER Grid program and with the Puget Sound Energy company, as well as building permit information and costs from Whatcom County. Following the analysis, the reader finds a discussion of the background steps, the proposal, the analysis, what the future of this proposal will look like, and how the reader may research their own personal energy system for their family.

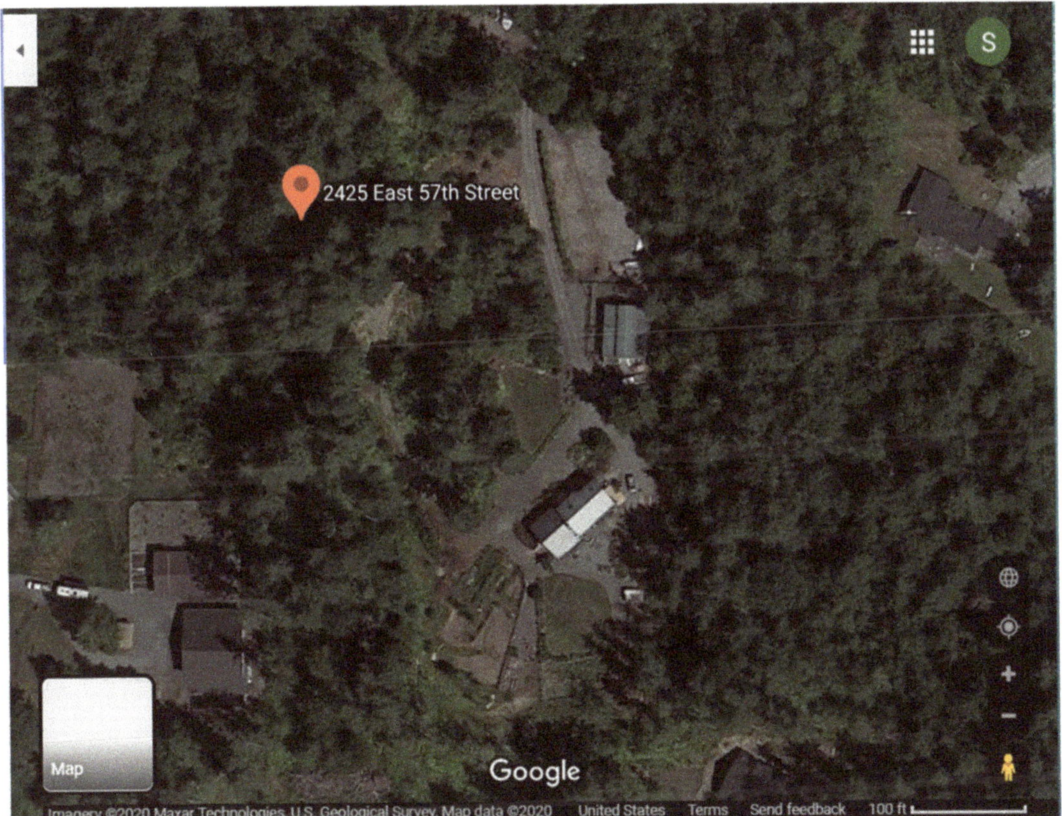

Figure 3. A screenshot of the plot of land and home for the wind turbine proposal (Google Maps).

3. Results

The project was set in Bellingham, Washington, a desirable living location for those searching for cooler climates, liberal policies, and a beautiful scenery. To begin, a home for sale was found using the website Zillow. This address was used in the HOMER Grid. The program prompts the user to download data from NASA.gov for wind speeds, radiation, and utility prices. The user must input the amount of electricity that they intend to use per day. The U.S. Energy Information Administration claims that most households in the country will use around 10,000 kWh per year, but the highest use is closer to 15,000 kWh per year. To ensure a fair outcome and some cushion for error, a higher energy use was assumed. The National Renewable Energy Laboratory recommends only a 1.5 kW wind turbine along with a solar panel to supplement. Wind turbines and solar panels create an optimum duo, since wind turbines create maximum electricity when solar panels are creating none [32].

The radiation in Bellingham appeared high and steady throughout, from one to six kWh/meter squared/day. The wind speeds appeared low, namely, under four meters per second throughout the year, as you can see in Figure 4. It appeared from the HOMER Grid resource projections that the house would benefit most from solar panels. Surprisingly, the program results showed otherwise. The solar panels would end up costing the user hundreds of thousands to millions of dollars more than the wind turbine that HOMER favored. The best result included using a specific 100-kW XANT wind turbine [31].

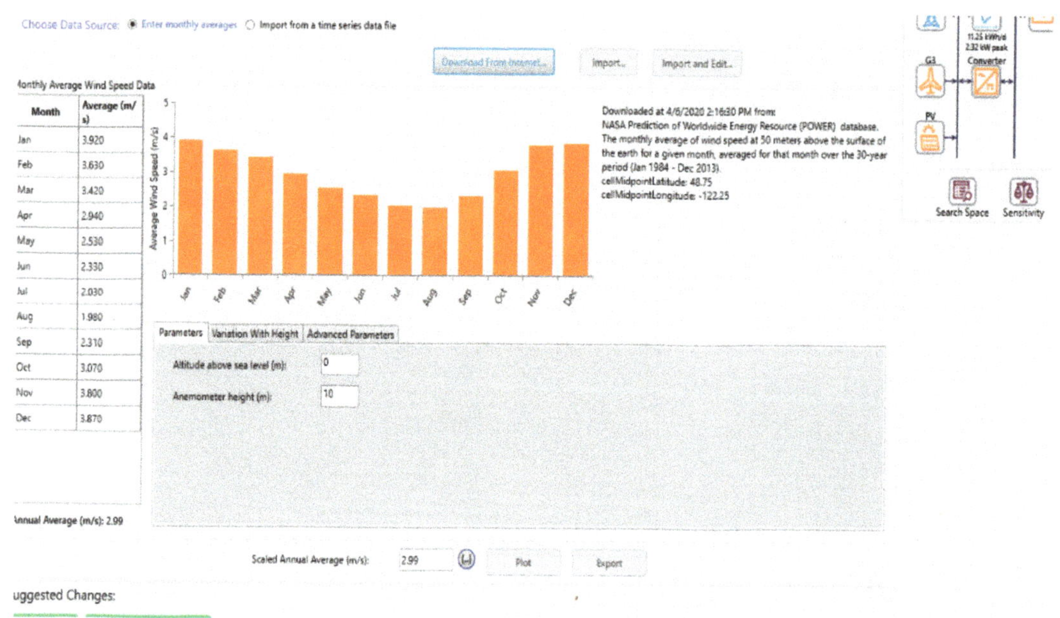

Figure 4. Screenshot of the HOMER Grid program with the wind resource information for Bellingham, Washington from NASA.gov (year: 2020).

The wind turbine options available in the HOMER Grid program include: XANT M-21 (100 kW), the XANT M-24 (95 kW), and three generic wind turbines with varying electricity production capacities: 1 kW, 3 kW, 10 kW, and 1.5 MW. The generic models were created by the HOMER Energy team to represent standard wind turbines with the latest technology. The XANT systems were created by the Belgian company of the same name. These turbines could easily be some of the best on the market for residential or community use. According to their website, these turbines have been IEC 64100-1 and GL certified (XANT, 2020). The International Electrotechnical Commission (IEC) was created in 1906 to showcase the "International Standards for all electrical, electronic, and related technologies", a standard approved by many state and national governments [33]. The turbines were created for easy transport, as it fits into a 40-foot container, can be erected without cranes, and can withstand storms [31]. The project set in Bellingham, Washington would require a building permit before moving forward. The permit ensures a bureaucratically labor-intensive process. The county of Whatcom requires that any residential wind turbine is under 100 kW. In fact, the program recommends this size and the proposal will include a specific 100-kW wind turbine from XANT. After this first win, there are many other conditions and administrations to check your property before building can commence [34]. The checklist in Figure 5 shows that the resident must also pay a fee for the application, but hidden is also the sewer verification process, the cost of the building plans, and possible extra costs for the site plan and rainwater verification. However, that amount may only be around $2000 total, according to the spreadsheet that the county provides on the webpage (2020). This number is added to the cost of the turbine, reaching an initial capital cost of $76,000. The HOMER Grid chose the XANT M-21 100 kW turbine because it had the most potential to create revenue for the property and the lowest Net Present Cost [31].

Items Required For a Complete Application:

Bring all of the following with you to your appointment:

☐ Whatcom County Planning & Development Approved Screener Checklist
  - Including your Natural Resources Assessment Approval or All Items Required through the Natural Resources Assessment

☐ Completed Building Permit Application Form – 3 pages (included in this packet)

☐ **2** Complete Sets of Building Plans and (if determined by staff) Structural Engineering

☐ **3** Copies of Site Plan

☐ Washington State Energy Code Form (Prescriptive Zone 1 Worksheet)

☐ Public, Private or Rain Water Verification
  - This form is NOT required if your building project:
    - Does not include plumbing for potable water, or
    - Is a residential remodel or addition that does not add additional bedrooms or result in an increase of floor space of more than 50%

☐ Whatcom County Health Department Approved Septic Permit & Design or Sewer Verification
  - AND if an existing septic a current inspection report completed by a licensed O & M specialist.

☐ Copy of Most Current Deed

☐ Current Contractor's License Number or Owner Contractor Statement of Understanding

☐ Whatcom County Engineering Approved Revocable Encroachment Permit

☐ Agent Authorization Form (if you are an agent applying on behalf of the owner)

☐ Cash or Check (U.S. Funds) are accepted as payment for fees. You may also pay by debit card or credit card, in person only, at the Planning & Development Services Permit Center Counter.

Figure 5. Screenshot of the building permit requirements for Whatcom County [34].

The next step was to run the simulations via HOMER Grid, examining all different scenarios. The scope of this work was to investigate different hybrid setups with the XANT M-21 100 kW wind turbine to optimally confront the challenge of setting up a small-scale renewable energy system. Inevitably, a general methodology and HOMER tool was used, focusing on proposing hybrid systems in autonomous grids, taking into account energy demand, local weather conditions, and long-term energy planning. Figure 6 shows the options provided by the HOMER Grid program after finding resources and choosing which solar panels and wind turbines to use for the analysis. The first option in Figure 6 shows the cost of using the wind turbine with the utility company. The simulation shows that this option would turn a profit if the utility company buys the surplus of the electricity produced. This is dependent on the production of the XANT M-21 100 kW wind turbine, which can produce higher amounts of electricity than most wind turbines due to the proficiency of the turbine technology. The next row shows the option of using the wind turbine with the utility company during the low wind speed months in the summer and a 1-kVA energy storage battery. This is the second-best option for the project. The third-best option is in the third row, which would be to use only the utility company and buy power like everyone else in the neighborhood. The options after that mix solar, wind, and storage, but are not profitable or even cost effective.

VarioTrack VT-65 (kW)	VarioTrack VT-65-MPPT (kW)	M-21	LI ASM	24	Converter (kW)	NPC ($)	COE ($)	Operating cost ($/yr)	Initial capital ($)	Ren Frac (%)	Total Fuel (L/yr)	IRR (%)	Simple Payback (yr)	Utility Bill ($/y
		1		1		$20,362	$0.0236	-$4,304	$76,000	90.3	0	6.2	12	$6,080
		1	1	1	0.125	$21,363	$0.0248	-$4,278	$76,664	90.6	0	6.1	13	$6,076
						$22,961	$0.116	$1,776	$0.00	0	0			$0
				1	0.250	$23,485	$0.119	$1,762	$701.50	0	0			$34.40
680	4.00			1	4.00	$1.91M	$1.86	$795.55	$1.90M	96.4	0			$7,813
680	4.00			1	4.00	$1.91M	$6.37	$6,719	$1.83M	71.2	0			$1,890
680	4.00		2	1	5.50	$1.93M	$6.35	$6,792	$1.83M	72.8	0			$1,929
680	4.00		2	1	3.50	$1.92M	$1.91	$1,047	$1.90M	97.1	0			$7,684

Figure 6. Screenshot of the HOMER Grid program results in Bellingham, Washington (2020).

The option chosen for the project is the XANT wind turbine along with the utility company. This option has a Net Present Cost (NPC) of $20,362, a levelized cost of electricity (LCOE) at $0.0236, a capital cost of $76,000, and operating cost (O&M) at −$4300 a month. This NPC represents the total cost of installing, operating, and maintaining the energy source minus the revenues it makes over its 25-year lifetime [31]. To look at this number another way, after 25 years, the total electricity bill will be about $20,000. When divided by the years in use, that is about $800 a year, a steal in electricity spending. One last important benefit to installing a wind turbine on one's property is the Federal Tax Credit, which pays back the consumer 22% of the installation and capital cost of the renewable energy system. However, this tax incentive expires in 2020, but if extended, this could make the NPC closer to $0 for the project.

4. Discussion

The wind turbine results in HOMER projected that it would pay itself back within 12 years, an amazing feat for investing in renewable energy. After using the program to find the NPC of the XANT 100 kW turbine, contacting the local energy supplier was crucial for verification. Email correspondence revealed that the Puget Sound Energy (PSE) company would not buy electricity for $0.90/kWh, as was illustrated by the computer software. Instead, the company would purchase the electricity at "the market competitive price of around 2 or 3 cents per kWh" (Zachary, M., personal communication, 24 April 2020). Reevaluating the numbers with the new data revealed less enthusiastic results. The HOMER Grid program found the yearly cost of the 100 kWh XANT wind turbine to be −$4304.00. This assumes that the surplus of electricity would be sold at the rate of 9 cents/kWh and that the operating costs would be covered by the revenue. The new numbers would present a yearly cost of −$707 and an NPC closer to $65,273 [34].

After contacting a representative at the Puget Sound Energy company, there are new costs to consider which were not prompted by the HOMER Grid software. Connecting to the grid can cost tens of thousands of dollars and takes two years, later learned from Mr. Zachary of the PSE. As the project unfolded, new questions unfolded; XANT claims on their website that they deliver a product which costs half the normal installation cost because their wind turbines do not need cranes to be erected. These costs are relevant, of course, to any person's specific location and needs and can vary from only a few thousand dollars to half a million dollars, leaving doubt as to what the actual capital costs would be [14].

The potentially high costs of connecting to the grid greatly offset the payback period, especially after the new data from Mr. Zachary about their price for the electricity. The PSE will, however, connect someone to the grid for almost no cost if the resident agrees to use their "net metering" program. This program offers credits for the resident's electricity generated and "sold" back to PSE; then, when the resident needs electricity while the turbine is not producing, they can use their credits to buy their electricity. This option may be practical if a cheaper wind turbine is purchased, such as the 1.5 to 3 kW, which are much cheaper. However, the HOMER Grid program chose the XANT M-21 100 kW wind turbine because of its wind energy capturing potential [35,36]. Therefore, the importance of other parameters is not taken into account—nor simulated in any way—and their impact might be decisive.

The importance of investing in renewable energy, such as a wind turbine, has proven essential throughout the last decade. Private citizens are even taking the initiative to buy solar panels for their roofs now that the technology is more affordable. Thanks to software such as HOMER Grid, communities and families can investigate their own home solar panel and/or wind turbine. Using the program, one can look up the resources in their area, such as radiance from the sun, wind speeds, and local utility company prices. After conducting the project, the authors recommend that future endeavors contact their utility company for information about buy back prices and grid connection to assure HOMER is accurate. However, it is not panacea. Calhoun et al. [37] have pointed out the possibilities of deception in simulations, pointing out the importance decision makers give to simulation results, while in another study, they stressed the educational and ethical implications [38]. Goldberg et al. [39] pointed out that deceptions of simulations should be considered as unexpected events and take a decision if these are real or not. Realism and a critical approach are always needed, especially when investing is part of the plan. Only few studies have focused on the pitfalls of the software tools, the deceptions in simulations, and the need to quantify decisive parameters that are usually omitted. Using HOMER in an example of a simple case in the US proves that random and qualitative parameters could make an investment from absolutely non-viable to viable and successful investment even if the simulation results are exactly the same. Therefore, it would be helpful to also contact others in the area who have followed through with their own wind turbine and/or solar panel for actual installation costs.

While the resources in the desired location in Bellingham, Washington were not promising, there was still a turbine on the market that could create more than enough electricity for the household. The building permit and the inspections on the property add time and additional cost capital costs. Moreover, connecting to the grid after the permit has been approved and the new turbine shipped and installed could take years to finish and possibly a few tens of thousands of dollars extra. All this time and money added along with the research time means it would take between two to five years to complete. It appears that the biggest obstacle for a consumer looking to lower their carbon footprint are the installation and grid connection costs as well as the loss of the Federal Tax Credit, which would rebate residents 22% of installation and operating costs—a huge benefit that needs to be renewed in 2020 before it expires [14].

In fact, the importance of making personal wind turbines easier and cheaper for residents and communities could be the pathway to relying 100% on renewables, similar to the case study of Denmark, a country currently relying on renewable energy for more than 50% of their energy needs. A movement by residents buying and investing in wind turbine shares drove research and development as well as anti-nuclear lobbyists, who pushed representatives to choose wind turbine investment over nuclear power in the 1970s [2]. While the US is littered with fossil fuel lobbyists, grassroots initiatives from homeowners to purchase wind power could be the accelerator this country needs right now. The nation's leaders need to cover grid connection costs and renew the federal tax credit to make owning personal wind turbines the future.

5. Conclusions

Wind and solar energy contributed in total more than 20% to Europe's power supply in the first half of 2020; Denmark reached 64%, followed by Ireland and Germany, with 49% and 42%, respectively. Thus, great achievements have been made for some countries, but there is still a long way to go, and small-scale wind energy is needed for achieving greater numbers for some countries, and a kick-start for others. This study focused on small-scale wind energy, and specifically on (a) the need of a generation to move on in producing their own electricity, either as a philosophy of life or as a status symbol and (b) the decision-making process. For the latter, it was revealed that there are tools that can simulate and propose hybrid system solutions; however, installation costs, maintenance costs, net metering options, and taxation schemes are not included in most of these tools. This is happening not only for this specific tool, but many more, especially in tools that are linked to investment decisions [40]. Therefore, it is meaningless and careless to only rely on the results of the tool and assume most of the above-mentioned parameters. The different assumptions in these parameters can make a project have a payback period from 10 to more than 20 years, or else from being a totally sustainable project to an inviable one.

However, when seen from a more general perspective, the practical cost for decision makers of counting only on simulation results can be unbearable when this is not coupled with the overall picture. A future educational agenda around the deception of simulations can advance the scientific areas of simulation as a learning tool.

Author Contributions: S.R. focused on data curation, formal analysis, methodology, software, and the writing of the original draft; G.X. focused on supervision and editing of the original draft, and performed the project administration. All authors have read and agreed to the published version of the manuscript.

Funding: This research received no external funding.

Institutional Review Board Statement: Not applicable.

Informed Consent Statement: Not applicable.

Data Availability Statement: The data presented in this study are available on request from the corresponding author.

Conflicts of Interest: The authors declare no conflict of interest.

References

1. Espinoza, A. Visitors get glimpse of Block Island Wind Farm in Test Phase. 2016. Available online: https://www.wnpr.org/post/visitors-get-glimpse-block-island-wind-farm-test-phase (accessed on 30 April 2020).
2. Valentine, S. *Wind Power Politics and Policy*; Oxford University Press: Oxford, UK, 2015.
3. Rasmussen, N.B.; Enevoldsen, P.; Xydis, G. Transformative multi-value business models: A bottom-up perspective on the hydrogen-based green transition for modern wind power cooperatives. *Int. J. Energy Res.* **2020**, *44*, 3990–4007. [CrossRef]
4. Koscis, G.; Xydis, G. Repair process analysis for Wind Turbines equipped with Hydraulic Pitch mechanism on the U.S. market in focus of cost optimization. *Appl. Sci.* **2019**, *9*, 3230. [CrossRef]
5. Wang, H.-M.S.; Piccard, L.; Yao, L.; Spohn, K.M. 2010, Feasibility study of wind power generation system at arctic valley. *EMJ Eng. Manag. J.* **2010**, *22*, 21–33.
6. Bastian, N.; Trainor, T. Going Green at West Point: Is It Economically Beneficial? A Cost-Benefit Analysis of Installing a Wind Farm at the United States Military Academy. *Eng. Manag. J.* **2010**, *22*, 12–20. [CrossRef]
7. Schoolman, E.D.; Shriberg, M.; Schwimmer, S.; Tysman, M. Green cities and ivory towers: How do higher education sustainability initiatives shape millennials' consumption practices? *J. Environ. Stud. Sci.* **2016**, *6*, 490–502. [CrossRef]
8. Nanaki, E.A.; Kiartzis, S.; Xydis, G.A. Are only demand-based policy incentives enough to deploy electromobility? *Policy Stud.* **2020**. In Press. [CrossRef]
9. Xydis, G.; Mihet-Popa, L. Wind Energy Integration via Residential Appliances. *Energy Effic.* **2016**, *10*, 319–329. [CrossRef]
10. Haynes, D.; Corns, S. Improving Grid Network Operations Through an Improved Energy Market. *EMJ Eng. Manag. J.* **2020**, *32*, 208–218. [CrossRef]

11. U.S. Energy Information Administration. How Much of U.S. Energy Consumption and Electricity Generation Comes from Renewable Energy Sources? 2018. Available online: https://www.eia.gov/tools/faqs/faq.php?id=92&t=4 (accessed on 19 April 2020).
12. Panagiotidis, P.; Effraimis, A.; Xydis, G. An R-focused Forecasting Approach for Efficient Demand Response Strategies in Autonomous Micro Grids. *Energy Environ.* **2019**, *30*, 63–80. [CrossRef]
13. Enevoldsen, P.; Xydis, G. Examining the trends of 35 years growth of key wind turbine components. *Energy Sustain. Dev.* **2019**, *50*, 18–26. [CrossRef]
14. Energy.gov. Small Wind Electric Systems. 2020. Available online: https://www.energy.gov/energysaver/save-electricity-and-fuel/buying-and-making-electricity/small-wind-electric-systems (accessed on 18 September 2020).
15. Hansen, J.M.; Xydis, G. Rural Electrification in Kenya. A useful case for remote areas in Sub-Saharan Africa. *Energy Effic.* **2020**, *13*, 257–272. [CrossRef]
16. Olasunkanmi, O.G.; Roleola, O.A.; Alao, P.O.; Oyedeji, O.; Onaifo, F. Hybridization energy systems for a rural area in Nigeria. *IOP Conf. Ser. Earth Environ. Sci.* **2019**, *331*, 012007. [CrossRef]
17. Suman, G.K.; Roy, O.P. 2019, Microgrid System for A Rural Area—An Analysis of HOMER Optimised Model Using MATLAB. In Proceedings of the 2019 3rd International Conference on Recent Developments in Control, Automation and Power Engineering, RDCAPE, Noida, India, 10–11 October 2019; pp. 534–539.
18. Hadjidj, M.S.; Bibi-Triki, N.; Didi, F. Analysis of the reliability of photovoltaic-micro-wind based hybrid power system with battery storage for optimized electricity generation at Tlemcen, north west Algeria. *Arch. Thermodyn.* **2019**, *40*, 161–185.
19. Oliver, D.; Groulx, D. Thermo-economic assessment of end user value in home and community scale renewable energy systems. *J. Renew. Sustain. Energy* **2012**, *4*, 023117. [CrossRef]
20. Ugur, E.; Elma, O.; Selamogullari, U.S.; Tanrioven, M.; Uzunoglu, M. 2013, Financial payback analysis of small wind turbines for a smart home application in Istanbul/Turkey. In Proceedings of the 2013 International Conference on Renewable Energy Research and Applications, ICRERA, Madrid, Spain, 20–23 October 2013; pp. 686–689.
21. Rodriguez-Hernandez, O.; Martinez, M.; Lopez-Villalobos, C.; Garcia, H.; Campos-Amezcua, R. Techno-economic feasibility study of small wind turbines in the Valley of Mexico metropolitan area. *Energies* **2019**, *12*, 890. [CrossRef]
22. Hemmati, R. Technical and economic analysis of home energy management system incorporating small-scale wind turbine and battery energy storage system. *J. Clean. Prod.* **2017**, *159*, 106–118. [CrossRef]
23. Canale, T.; Ismail, K.A.R.; Lino, F.A.M.; Arabkoohsar, A. Comparative Study of New Airfoils for Small Horizontal Axis Wind Turbines. *J. Solar Energy Eng.* **2020**, *142*. [CrossRef]
24. Olaofe, Z.O.; Folly, K.A. Potentials of a 5KW wind energy system with integrated storage bank for home energy management. In Proceedings of the IASTED International Conference, Botswana, Africa, 3–5 September 2012; pp. 102–109.
25. Rasouli, V.; Hemmati, R. Net zero energy home including photovoltaic solar cells, wind turbines, battery energy storage systems and hydrogen vehicles. In Proceedings of the 2017 International Conference in Energy and Sustainability in Small Developing Economies (ES2DE), Funchal, Portugal, 10–12 July 2017. [CrossRef]
26. Statista. Quarterly growth of the real GDP in the United States from 2011 to 2020. 2020. Available online: https://www.statista.com/statistics/188185/percent-change-from-preceding-period-in-real-gdp-in-the-us/ (accessed on 20 April 2020).
27. Keys, L.K. The late second coming of the automobile. *EMJ Eng. Manag. J.* **1993**, *5*, 11–16. [CrossRef]
28. Michalitsakos, P.; Mihet-Popa, L.; Xydis, G. A Hybrid RES Distributed Generation System for Autonomous Islands: A DER-CAM and Storage-Based Economic and Optimal Dispatch Analysis. *Sustainability* **2017**, *9*, 2010. [CrossRef]
29. Avgoustaki, D.; Xydis, G. Indoor Vertical Farming in the Urban Nexus Context: Business Growth and Resource Savings. *Sustainability* **2020**, *12*, 1965. [CrossRef]
30. Egilmez, G.; Oztanriseven, F.; Gedik, R. The energy climate water nexus: A global sustainability impact assessment of U.S. manufacturing. *Eng. Manag. J.* **2020**, *32*, 298–315. [CrossRef]
31. HOMER Grid. "1.7.7394.26541". Available online: https://www.homerenergy.com/products/grid/index.html (accessed on 4 April 2020).
32. National Renewable Energy Laboratory. What Size Wind Turbine Do I Need? OpenEI.com. 2016. Available online: https://openei.org/wiki/Small_Wind_Guidebook/What_Size_Wind_Turbine_Do_I_Need (accessed on 3 April 2020).
33. International Electrotechnical Commission. "About the IEC," iec.ch. 2020. Available online: https://www.iec.ch/about/?ref=menu (accessed on 23 April 2020).
34. Personius, M. "Planning and Development Services," Whatcomcounty.us. 2020. Available online: https://www.whatcomcounty.us/861/Permit-Center (accessed on 1 April 2020).
35. Puget Sound Energy. Distributed Renewables. 2020. Available online: https://www.pse.com/green-options/Renewable-Energy-Programs/distributed-renewables (accessed on 10 April 2020).
36. Nazir, M.S.; Wang, Y.; Bilal, M.; Sohail, H.M.; Kadhem, A.A.; Nazir, H.M.R.; Abdalla, A.N.; Ma, Y. Comparison of Small-Scale Wind Energy Conversion Systems: Economic Indexes. *Clean Technol.* **2020**, *2*, 144–155. [CrossRef]
37. Calhoun, A.; Pian-Smith, M.; Shah, A.; Levine, A.; Gaba, D.; DeMaria, S.; Goldberg, A.; Meyer, E.C. Exploring the boundaries of deception in simulation: A mixed-methods study. *Clin. Simul. Nurs.* **2020**, *40*, 7–16. [CrossRef]
38. Calhoun, A.W.; Pian-Smith, M.C.; Truog, R.D.; Gaba, D.M.; Meyer, E.C. Deception and simulation education: Issues, concepts, and commentary. *Simul. Healthc.* **2015**, *10*, 163–169. [CrossRef] [PubMed]

39. Goldberg, A.T.; Katz, D.; Levine, A.I.; Demaria, S. The importance of deception in simulation: An imperative to train in realism. *Simul. Healthc.* **2015**, *10*, 386–387. [CrossRef] [PubMed]
40. Xydis, G. A wind energy integration analysis using wind resource assessment as a decision tool for promoting sustainable energy utilization in agriculture. *J. Clean. Prod.* **2015**, *96*, 476–485. [CrossRef]

Article

Comparison of Different References When Assessing PV HC in Distribution Networks

Samar Fatima, Verner Püvi and Matti Lehtonen *

Department of Electrical Engineering and Automation, Aalto University, Maarintie 8, 02150 Espoo, Finland; samar.fatima@aalto.fi (S.F.); verner.puvi@aalto.fi (V.P.)
* Correspondence: matti.lehtonen@aalto.fi

Abstract: The burgeoning photovoltaics' (PVs) penetration in the low voltage distribution networks can cause operational bottlenecks if the PV integration exceeds the threshold known as hosting capacity (HC). There has been no common consensus on defining HC, and its numerical value varies depending on the reference used. Therefore, this article compared the HC values of three types of networks in rural, suburban, and urban regions for different HC reference definitions. The comparison was made under balanced and unbalanced PV deployment scenarios and also for two different network loading conditions. A Monte Carlo (MC) simulation approach was utilized to consider the intermittency of PV power and varying loading conditions. The stochastic analysis of the networks was implemented by carrying out a large number of simulation scenarios, which led towards the determination of the maximum amount of PV generation in each network case.

Keywords: distribution networks; Monte Carlo simulations; PV hosting capacity; photovoltaics

1. Introduction

The hosting capacity (HC) concept has been gaining importance with time to ensure the capacity of the system without employing any expensive grid upgrades. However, the value of HC varies considerably depending on a variety of factors including photovoltaic (PV) locations, network loading conditions, numerical values of limiting factors, and PV deployment scenarios. Moreover, HC value is dependent on the references used for its definition, and Reference [1] provides a review of different HC references and their influence on changing HC value. The study concluded the five major HC references used in the literature to be peak load, transformer rating, the share of customers' PVs, energy consumption, and share of available roof space. The HC can be defined as the ratio of maximum PV production to the peak load of the feeder [2] or transformer's kVA rating [3], or the ratio of total yearly PV production to the yearly energy consumption [4] and w.r.t roof space [5]. Alternatively, it can be defined with respect to the customers equipped with PVs as the ratio of customers with PVs to the total customers in the area under study [6]. The grid operators are concerned about maintaining the power quality standards with an increasing trend to integrate more PVs in the electrical networks by using novel technologies. The compliance with the performance constraints also known as the HC limiting factors is an important criterion for the accurate determination of HC without risking the quality of supply and the network component's life. The limiting factors for defining the HC are voltage variations, voltage unbalance, overloading limit of cables and transformers, flicker, and harmonics. Voltage rise is the important limiting factor in low-voltage networks in terms of voltage variations without any significant contribution of under-voltage in limiting the HC. Therefore, the under-voltage limit violation can be excluded from the set of limiting factors. The accurate choice of the limiting factors and their operational threshold significantly influence the HC of the network, and different studies have used a variety of performance indices for HC determination. Overvoltage limit

Citation: Fatima, S.; Püvi, V.; Lehtonen, M. Comparison of Different References When Assessing PV HC in Distribution Networks. *Clean Technol.* **2021**, *3*, 123–137. https://doi.org/10.3390/cleantechnol3010008

Received: 5 January 2021
Accepted: 27 January 2021
Published: 1 February 2021

Publisher's Note: MDPI stays neutral with regard to jurisdictional claims in published maps and institutional affiliations.

has been widely used in the literature as a performance constraint followed by overloading limits of cables and transformers.

The study conducted in [2] focused on finding the technical constraints limiting the PV HC and concluded that on-load tap changer (OLTC) is more effective in HC improvement in balanced PV deployment as compared to the single-phase connections. A Monte Carlo (MC)-based simulation analysis was carried out in [3], considering PV allocation as a random variable, with the authors discussing the HC dependence on the loading level of the network. According to this study, 0–5% loading enabled only 10% of investigated LV (low voltage) circuits to host 30% PV capacity as compared to the 80% LV systems with midday loading of 25–35% to host similar PVs. However, the authors argued that PV hosting capacity is not merely dependent on the loading levels but also on other factors such as the number of customers, length of the LV network, PV connection scenarios, and load distribution of different types of customers. The idea of HC dependence on varying factors is substantiated by a study performed in [7]. The authors here discussed that the prediction of HC for any network is subjected to many uncertainties: variable solar production, customer installations, PV connections such as single-phase or three-phase, and panel tilt, among many others. Similarly, two different HC values of 43% and 83% w.r.t energy consumption were investigated in [4], considering the very well suited rooftops and all rooftops for PV connection, respectively. Moreover, the potential of the battery energy storage system, active power curtailment, and dynamic thermal loading of the transformer was investigated in this study to increase the PV HC. However, the dynamic loading of the transformer as a potential means to improve HC was found to be effective only in the case of the transformer that is loaded for the short term.

The quantification of rooftop solar PV potential plays a vital role in the HC calculation w.r.t roof space. This quantification involves the estimation of available rooftop area for the installation of solar PV panels [8] as the entire roof area cannot be utilized for this purpose due to varying reasons such as shadows from surrounding buildings and the mechanical barriers such as ventilation equipment, etc. A similar form of research was conducted in [9] that was based on the calculation of usable rooftop area for solar PV installation as part of the Energizing Urban Ecosystems project founded by RYM Oy using realistic solar radiation data in the city of Espoo, Finland. The author utilized three filters: profitability filter, city planning filter, and mechanical barrier filter to eliminate the unsuitable areas for solar PV installations. The HC calculation w.r.t roofspace PVs was carried out in [5], wherein the authors employed a model predictive control strategy for the determination of HC. The authors investigated six reference grids in the remote, rural, and urban regions with HC values of 16%, 13%, and 45% w.r.t roof space, respectively. The effectiveness of power curtailment in conjunction with the storage options and reactive power control is also discussed in this study to increase the PV penetration and thus the network HC. The authors in this study further noticed that HC was limited by voltage violations in the rural and remote networks, whereas the urban grid's HC was restricted by the thermal violations.

A real UK residential network with single-phase customers was analyzed in [6] to determine the HC value w.r.t customers with PVs, and the potential of OLTC with a setting of $+/-8\%$ for HC improvement was investigated. An MC-based approach was employed by the authors in this work to take into consideration the uncertainties in the load and PV generation. Similarly, a static load type feeder was investigated in [10] for the HC determination, considering voltage, power flow, and cable ampacity as the limiting factors, with HC linearly related to the load variations. In this case, there was also reverse power limitation, and the authors noticed this kind of power flow violations at the load value of 8.8 MW, which is the current load of an actual 20 kV distribution feeder in an urban distribution network with the attribute of large loads. Moreover, the HC was also defined in terms of the actual active power of the load in [11], where the authors investigated an Australian distribution feeder in the context of validation of the voltage rise mitigation strategy proposed in the study. The calculation approach for the HC w.r.t active power was similar to the HC w.r.t peak load. However, the authors calculated the HC values at two different times of the day, morning and midday, and found that the midday loading enabled more HC as compared to morning loading. The aggregated loads on the distribution feeder were modelled by considering the characteristic electrical appliances of the residential household, and the PV output power was compared with the actual active power of the loads at the point of common coupling. The study revealed the HC value of the network as 111% with PV output power as 1.73 kW while serving a load of 1.5 kW at 9 a.m. The HC value of this network was increased to a value of 177% at midday, with an increased PV power of 2.93 kW serving the loading level of 1.65 kW at the point of common coupling. Thus, the findings of the study revealed the higher value of HC at midday as compared to morning, which was due to increased load. The accurate HC determination enabled the utilities to make timely decisions to integrate or curtail the future PV penetration to ensure the reliability of electrical network. The idea of changing HC due to different reference definition can be further explored by finding the gap between the HC values of a network by using different references. The core idea of this study was to compare different references used when determining numerical values of PV HC in distribution networks. The main purpose of this study was to show that HC value of the same network can vary depending on different references and how these values are far apart.

The HC determination should be carefully done by either a well-informed guess [7] or a stochastic approach, and thus this study focused on the latter approach. The first part of the HC calculation modified and adopted an MC simulation-based approach developed in [2] that was further extended to find the HC values w.r.t different references. Additionally, the impact of variation in HC references was analyzed under balanced and unbalanced PV connection scenarios. The value of HC w.r.t peak load or energy consumption is largely dependent on the loading of the network, and thus the adopted model was simulated by randomly sampling the load values among three types of loads. The customer loading data were primarily based on the heating modes for three distinct Finnish networks in rural, suburban, and urban regions. Moreover, HC values are also influenced by the loading level; therefore, this aspect of network loading was taken into account in the analysis of PV HC by taking maximum load as 100% and 50% of transformer rating, respectively. An hourly stochastic analysis was utilized for determining the network HC using the hourly data, showing variations in load demand and PV generation for an entire year. Moreover, the proposed hourly MC-based algorithm employed the approach of the worst-case hours that are of high concern for the network planning to ensure network capacity even under the highest PV penetration during these hours. This aspect facilitated the network planners responsible for making crucial decisions regarding network reinforcement investments. An adequately high number of MC simulations was carried out for the accuracy of HC results by simulating a large number of iterations. Finally, a sensitivity analysis was carried out for two loading levels at three different load power factor settings to validate the HC values of the networks.

This paper is organized as follows. Section 2 introduces the system models; the nodes; and customers per feeders of the networks, source data, and the limiting constraints for PV HC. Section 3 begins by introducing the assessment methodology, loading contribution based on heating modes, and the explanation of the proposed MC algorithm. Section 4 defines the different HC w.r.t references and calculates the HC of the networks under varying scenarios with respect to five HC references: peak load, transformer rating, energy consumption, the ratio of customers with PVs, and ratio of roof-space used for PVs. Thus, this work can serve as a foundation to maintain the fact that HC of a network is not a unique value. Section 5 provides a summary of the results.

2. Simulation Characterization and Source Data

The investigated LV networks of this study are formulated in [2], comprising Finnish predominantly rural (PR), suburban (SU), and predominantly urban (PU) regions with different proportions of customers per node, as shown in Table 1. The customers were distributed homogenously among the nodes and PV distribution was tested for both the balanced and unbalanced feed-in cases. The following limiting factors for HC analysis were employed in this study:

- Upper voltage limit as +5% of Un (E1).
- Static loading of cables (E2).
- Transformer static overloading limit (E3).
- Negative sequence voltage unbalance limit as 2% of Un (E4).
- Ampacity limit of neutral-wire (E5).

Table 1. LV test networks' characteristics and grid components specifications.

Region	No. of Feeders	Nodes/Feeder	Customers/Node	Total Customers	Transformer (kVA)	Cable Length (m)
PR	1	8	1	8	50	150
SU	3	4,3,3	4	40	200	100
PU	3	2,2,1	60	300	1000	100

The nature of loads influences the determination of HC, and the loads in the LV systems are higher in resistive components. Therefore, a higher power factor (0.95) [2] and constant power loads were simulated in this study. The simulation assumptions are discussed further in Section 4.

In addition, the solar input data were based on the research conducted in [12], which provided the simulation of PV generation stochastic variation over an entire calendar year. The load data utilized in the simulation model consisted of three types of customers for an entire year on an hourly resolution. The load consumption profiles were based on the heating modes of the customers, as given in Table 2. The different modes of heating are defined as storage heating, district heating, and dielectric heating. The storage heating is the form of electrical heating with heat storage such as an electric boiler that is usually charged at night-time, thus utilizing the benefit of the reduced night-time tariffs. In contrast, the district heating also represents the houses with some other heating method such as fuels, including but not limited to oil, natural gas, and wood. Moreover, these customers do not utilize the electrical heating system for space heating. The annual load profiles were generated by incorporating the information about the number of customers and their heating methods, and finally the peak load of the feeder was calculated for further analysis. However, the peak load was then scaled according to the feeding transformer capacity using a scaling factor. The scaling factor employed here is defined as the ratio of transformer rating and the feeder load [13]. The scaling factor and different network loading will be further discussed in Section 4. Similarly, the PV generation data were collected for the complete year in the form of theoretical maximum PV generation without considering the weather conditions that affect irradiance profiles. The unbalance condition of naturally unbalanced systems can be increased further by the connection of single-phase PV installations. The lognormal distribution function of load unbalance data employed in this study was based on a single household in Helsinki, Finland, as used also in a study conducted by [14] that was further used for the determination of voltage unbalance magnitude and the angle.

Table 2. Loading (heating) distributions for three types of investigated regions.

Region	Storage Heating (%)	District Heating (%)	Direct Electric Heating (%)
PR	5.9	52.9	41.2
SU	7.6	52.5	39.9
PU	0.5	95.3	4.2

3. Assessment Methodology for PV HC Determination

A Monte Carlo-based algorithm was proposed for the determination of HC, considering its dependence on various operational scenarios. The load types depending on heating modes were randomly selected by running the MC simulations 1000 times [14] to ensure the accuracy of HC results. The changing loading profiles of each region were sampled randomly according to the percentages of three types of loads: storage heating, district heating, and direct electric heating, as given in Table 2. This load profile randomization was primarily performed to take into account the stochastic nature of the loads. The MC simulation model was based on the worst-case hours where the maximum PV generation coincided with the minimum network load consumption. The worst-case hours were selected out of a total of 8760 h for the load and PV values to reduce computational efforts based on the convex hull approach employed in [14]. The worst-case hours approach is considered a conservative approach. The random variables involved in this study were mainly the load profiles and PV deployment scenarios.

The general methodology used for hosting capacity determination is presented in Figure 1, the details of the proposed MC algorithm developed in MATLAB platform are shown in Figure 2a, and different steps are described below.

1. The model commences by defining the network parameters such as the number of nodes, impedance of lines and transformers, and the base PV defined as 1 kWp for each region.
2. This is followed by the main MC algorithm that increments the PV power of each PV module installed in the network. It starts with the first scenario by starting the PV generation at 1 kWp incremented in steps of 100 W until the maximum PV that is taken as equal to the rated transformer power of the region under consideration.
3. The loading profiles depending on the region selected in step 1 are randomly sampled at this stage. This algorithm simulates for 1000 different loading profiles for the accuracy of HC results. Similarly, this stage involves the random allocation of single and three-phase PVs on the nodes of the network. This step includes the selection of worst-case hours from the total 8760 h and finding the worst-case PV and loading scenario.
4. The power flow analysis based on backward and forward sweep (BFS) load flow analysis is simulated in a time-series framework as the loading profiles are changing with time. This is followed by checking the possible violations of performance constraints and the results are saved after checking the violations.
5. The scenario count is checked at this step and the model simulates again from step 3 if the scenario count is not reached at 1000 iterations. Alternatively, the constraints violation is checked and the PV size is incremented by a step of 100 W if violations are within 5% tolerance level of the total grid violations [2]. The model works such that the PV increment is discontinued when the performance constraint exceeds the predefined limits for more than 5% tolerance level. The idea behind setting the grid violations to be under 5% corresponds to the power quality requirements of supply voltage variations and supply voltage unbalance to be within 95% confidence limit as per the Standard EN 50160.

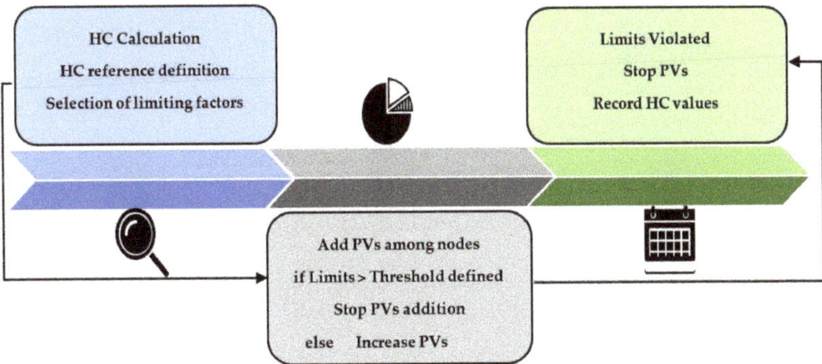

Figure 1. General methodology for hosting capacity (HC) determination.

This MC algorithm is a modified version of the method presented in [15] to calculate the HC w.r.t customers' PVs by incrementing the customers equipped with PVs until the grid violations are detected, as shown in Figure 2b. Steps 2–5 of the main algorithm in Figure 2a are modified such that the customers with the PVs are incremented in steps of 1 until the grid observes 5% of violations [2]. However, the random selection of load types and worst-case hour selection follows the same procedure as explained in step 3 by utilizing the same function in the MATLAB script. The load types and hence load values for one year were randomly sampled and the worst-case hours were estimated for a total of 1000 iterations to ensure the sufficient accuracy of HC results.

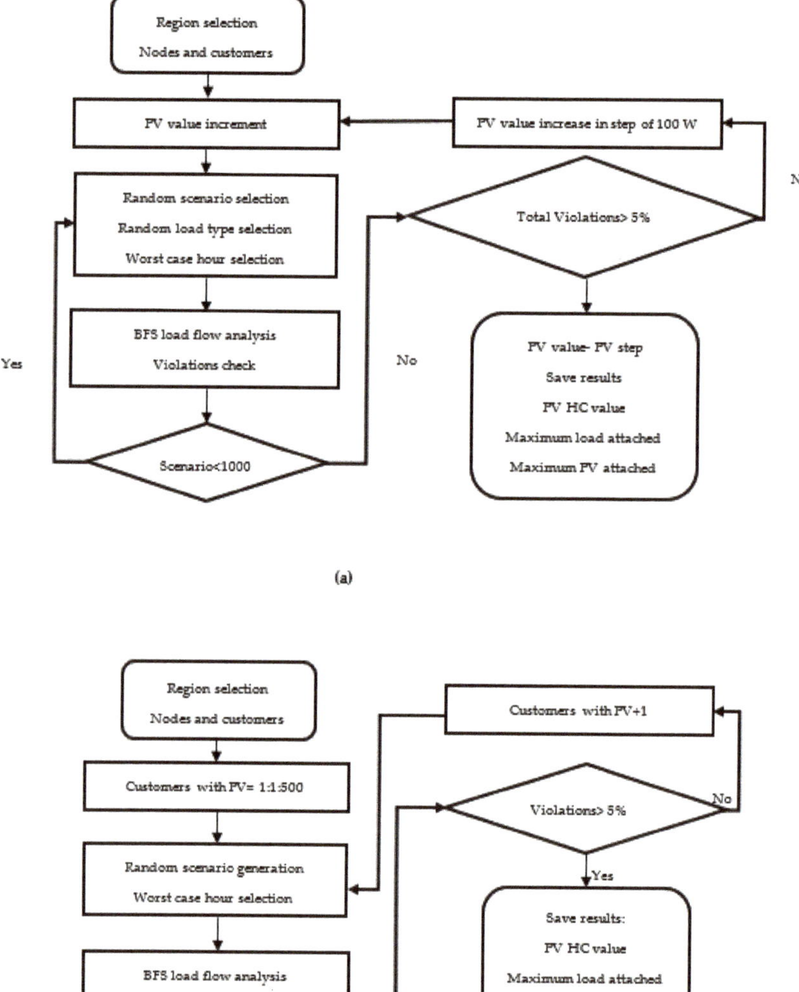

Figure 2. The Monte Carlo (MC) algorithm for the calculation of HC: (**a**) HC w.r.t peak load, transformer rating, energy consumption, and roof-space; (**b**) HC w.r.t customers' photovoltaics (PVs).

4. HC Analysis w.r.t Different References

PV HC can be defined in multiple ways, employing various references, as shown in Table 3. The HC of three regions—rural, suburban, and urban—were calculated in this section according to the references defined below. The simulations were performed for the three regions separately for a balanced and unbalanced PV feed-in scenario by first taking the maximum load as 100% of the transformer rating and then 50 percent of the transformer rating.

Table 3. Different definitions of HC w.r.t varying references [1].

HC Reference	HC Definition
Peak load	The ratio of peak PV capacity to the peak load of the feeder.
Transformer rating	The ratio of peak PV capacity to the transformer rating.
Energy consumption	The ratio of total yearly PV production to the total yearly energy consumption.
Customers' PVs	The ratio of customers with PVs to the total number of customers.
Roof space PVs	The ratio of roof space utilized for PV installation to total available roof space.

The HC research conducted in this part of the article is based on the following assumptions.

1. HC calculation w.r.t customers' PVs considers the increment of customers equipped with PVs in steps of 1 kWp for each penetration level as in the study performed in [15].
2. HC calculation w.r.t other references assume the even distribution of PVs along the length of the feeder with PVs at each bus. Moreover, the base PV for the rural, suburban, and urban regions is taken as 1 kW [2].
3. The HC w.r.t energy consumption is calculated assuming all the customers having PV installation as calculated in [4].
4. The proposed model takes into account only the theoretical maximum PV generation without considering the weather conditions impacting the irradiance level.
5. The HC is calculated for the load power factor as 0.95 [2].
6. The PV power factor is considered as unity.

The HC w.r.t roof-space was also defined as the roof area of the customers connected with the feeder that potentially enables the installation of PV panels. The determination of the usable rooftop area for PV installation is an extensive task that requires statistical data of the actual buildings by first finding the total roof area of the region to be studied and then finding the usable rooftop area for PV installation. Therefore, HC w.r.t roof space was calculated in this work by first estimating the practical roof space employed for potential PV installation. A PV panel of the power range between 260 and 285 Wp was utilized in this study with a size of 1.6 m^2 [16]. Hence, a PV array consisting of four PV panels of the power range 260 W was used for generating 1 kWp PV output covering an area of 6.4 m^2. The analysis then further proceeds towards finding the rooftop area to install the PV panels that satisfy the maximum PV generation during the worst-case hours. The estimation of the total available usable roof space area for PV installation is based on the research conducted in [9]. The HC w.r.t roof space was finally calculated as the ratio between the roof space area utilized for maximum PV generation during worst-case hours and the total available roof area for PV installation after eliminating the unsuitable areas.

4.1. Case Studies and Results

This section presents the HC analysis of the three geographically distinct Finnish regions according to different HC references for two types of PV deployment scenarios. The following scenarios are considered for HC analysis.

- HC analysis without scaling maximum load.
- HC analysis by scaling maximum load as 100% and 50% of transformer rating.

4.1.1. HC Analysis without Scaling Maximum Load

In this part of study, the LV networks were first analyzed depending on the loading level in comparison to the transformer rating, as given in Table 4. It was observed that the network load was only about 35–55% of the transformer rating, such that the transformers were not fully loaded initially and thus leaving sufficient headroom for PV addition without risking the transformers' life. Moreover, this part of the work utilized the original peak load value without scaling the peak load w.r.t transformer rating that would be discussed further in Section 4.1.2. The LV system loads were predominantly resistive and they were modeled by considering a high-power factor (0.95). Moreover, the constant power loads are depicted in Equation (1) according to a study conducted in [2] as active and reactive power profiles. The right-hand side of Equation (1) represents the real and reactive powers and the left side represents the apparent power of the constant power loads.

$$[S_L^{abc}] = [P_{PQ}^{abc}] + i[Q_{PQ}^{abc}] \tag{1}$$

Table 4. Average load values of the networks under balanced PVs in three regions for worst-case hours (minimum load values) and peak load calculated among all the 8760 h.

Region	Worst-Case Hour Load Values (kW)	Peak Load (kW)	Transformer Rating (kVA)	Initial Peak Load/TF Rating (%)
PR	5.9	20.8	50	41.6
SU	31	109.5	200	54.7
PU	82.6	359.7	1000	35.9

The load modeling for the calculation of PV HC was highly dependent on the type of heating mode used for each region, as described in Table 2. The load values depending on the heating mode of each region were first randomly sampled by incorporating a random probability distribution function "randp" in MATLAB to create a vector of the size of "number of network nodes X 1". This column vector of the size of network nodes was then utilized further to calculate the load values for the complete year by using the load distribution data among three types of customers utilizing different heating modes. Finally, a matrix of the size of "number of network nodes X 8760" was formed, showing the load values at each node for one complete year (8760 h). Afterwards, the peak feeder load was calculated by aggregating the load values among all the nodes connected to a feeder for 8760 h (a matrix of 1 × 8760), and then the peak load was selected as the maximum value from this row vector of the size 8760. Thus, the loading data were based on realistic load values by using real consumption profiles of the customers.

The load values utilized in this work were basically of two types: the minimum load value corresponding to the worst-case hours and the peak load value calculated by finding the maximum feeder load value from 8760 h. Moreover, the HC determination w.r.t peak load and energy consumption involved the peak feeder load instead of the worst-case hour load value. The peak load and the mean load value among the worst-case hours are given in Table 4. An MC-based simulation analysis was performed in this section for investigation of the HC w.r.t different references for two PV deployment scenarios: balanced 3-phase PV and unbalanced 1-phase PV, as presented in Table 5, and also in Figure 3. The HC results show that the HC of the rural region was the lowest among the three regions and that the balanced PV scenario permitted higher values of HC as compared to an unbalanced PV connection.

The investigation of the limiting constraints of the HC shows that the overvoltage limit restricted the PV HC in the balanced rural networks and the transformer loading limited the PV penetration in the suburban and urban regions. Moreover, the negative sequence unbalance remained the limiting constraint in the rural and suburban unbalanced PV connection scenarios and neutral wire ampacity limited the PV in urban unbalanced PV deployment. The sequence of the limiting factors such as E5, E4, and E2 in the unbalanced urban network represented the frequency of occurrences of the violation of limiting factors. The results revealed that the cable ampacity manifested itself as a limiting factor for HC only in the unbalanced urban networks and its violation occurrences were outnumbered by the other two limiting factors of neutral wire ampacity and voltage unbalance. Moreover, a higher value of PV HC w.r.t peak load of the urban region in Table 5 can be attributed to the initial peak load of urban region to be only about 35.9% of the transformer rating. Thus, this lower value of peak load in the denominator of HC definition as given in Equation (2) of Section 4.1.2 resulted in a higher value of HC.

Table 5. HC values as compared to different references for balanced and unbalanced PV scenarios.

HC References	Balanced (Three-Phase PV)		Unbalanced (One-Phase PV)	
Rural	**HC (%)**	**Violation**	**HC (%)**	**Violation**
Peak load	148	E1 [1]	65.6	E4, E1
Transformer rating	55.6	E1	26	E4, E1
Energy consumption	91.7	E1	42	E4, E1
Customers' PVs	50	E1	25	E4, E1
Roof-space PVs	8	E1	3.7	E4, E1
Suburban				
Peak load	219	E3	103	E4
Transformer rating	110	E3	60	E4
Energy consumption	135	E3	67.9	E4
Customers' PVs	57.5	E3	35	E4
Roof-space PVs	12.5	E3	6.8	E4
Urban				
Peak load	301	E3	108	E5, E4, E2
Transformer rating	107.8	E3	40.7	E5, E4, E2
Energy consumption	248	E3	85	E5, E4, E2
Customers' PVs	73.6	E3	26	E5
Roof-space PVs	8.2	E3	3.1	E5, E4, E2

[1] Over-voltage (E1), cable ampacity (E2), transformer overloading (E3), voltage unbalance (E4), and neutral-wire ampacity (E5).

This study adopted a more conservative voltage rise limit of +5% of the nominal voltage, resulting in the lower value of HC with respect to transformer rating of 55% for rural region under balanced PV scenario as compared to the HC (105%) in [14]. This is attributed to the different selection of performance constraints, and therefore it led towards the fact of how HC values of similar networks can be altered by choosing different limits of the performance constraints. However, this change in voltage rise constraint impacted merely the HC of the rural region where the HC was primarily limited by over-voltage as compared to the transformer overloading limiting the HC of suburban and urban regions.

4.1.2. HC Analysis by Scaling Maximum Load as 100% and 50% of Transformer Rating

The maximum PV HC was strongly influenced by network loading. This part of the work was focused on the determination of the impacts of variation of maximum loading of the network on the PV HC of the network. The HC results for two loading levels are given in Table 6. This analysis shows how different limiting factors of the HC manifested themselves under varying loading conditions, thus changing the HC of the network.

The calculation of HC w.r.t two loadings levels was initiated by generating the hourly annual load profiles for a year, as described in Section 3. The loading profiles were then scaled by using a scaling factor for 100% and 50% of the transformer rating. The annual hourly loading profiles at each node in the network were used for estimating the maximum feeder load of the network that was compared with the 100% and 50% of the transformer rating. Thus, the comparison of the peak load with the transformer rating revealed the lower value of maximum load, and thus the maximum load of the network was scaled up by using a scaling factor. The selection of the scaling factor played a central role in the accurate determination of the HC of the network and it is defined here as the ratio of the transformer apparent power to the peak feeder load. The apparent power of the transformers was distinct for each region, and therefore the scaling factor varied for each of the investigated regions. Similarly, the scaling factor for 100% and 50% of the transformer rating was different, leading to the different values of PV HC. The research results depicted that HC values w.r.t peak load and energy consumption were highly skewed by almost doubling in magnitude with maximum load as 50% of the transformer rating as compared to 100% of the transformer rating.

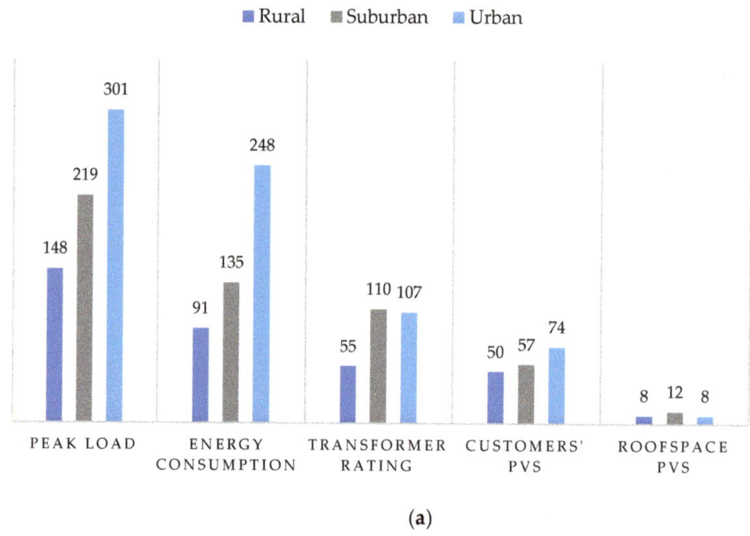

(**a**)

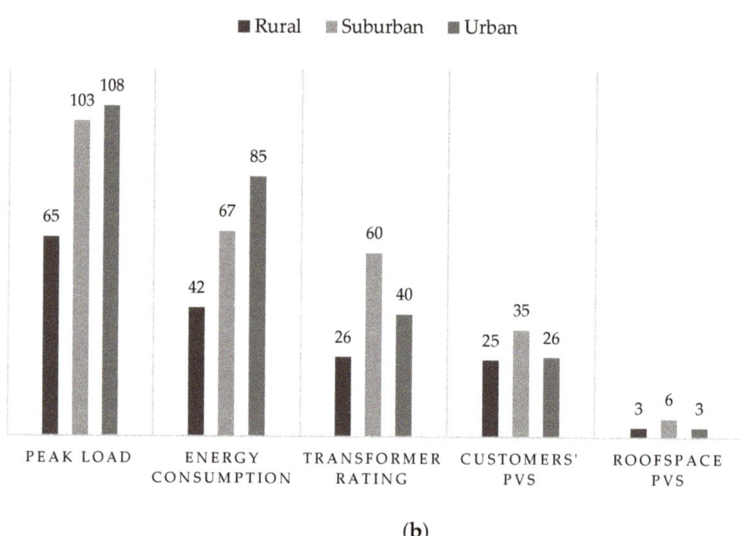

(**b**)

Figure 3. The comparison of HC w.r.t different references: (**a**) HC comparison for balanced PV deployment; (**b**) HC comparison for unbalanced PV deployment.

Table 6. HC reference value comparison for two different loading levels of the maximum load as 100% and 50% of transformer rating. (**a**) HC of the rural region for two loading levels; (**b**) HC of the suburban region for two loading levels; (**c**) HC of the urban region for two loading levels.

(**a**)

Loading Level	100% of Transformer Rating				50% of Transformer Rating			
PV Connection	Balanced		Unbalanced		Balanced		Unbalanced	
HC References	HC	Limit	HC	Limit	HC	Limit	HC	Limit
Peak load	74	E1	26	E4	120	E1	55.6	E4, E1
Energy consumption	45	E1	18	E4	74	E1	35.5	E4, E1
Transformer rating	74	E1	26	E4	60	E1	27.8	E4, E1
Customers' PVs	62.5	E1	25	E4, E1	50	E1	25	E4, E1
Roof-space PVs	10.58	E1	3.7	E4	8.6	E1	3.9	E4, E1

(**b**)

Loading Level	100% of Transformer Rating				50% of Transformer Rating			
PV Connection	Balanced		Unbalanced		Balanced		Unbalanced	
HC References	HC	Limit	HC	Limit	HC	Limit	HC	Limit
Peak load	123.7	E3	58	E4	225	E3	120	E4
Energy consumption	80	E3	36.5	E4	140.5	E3	75	E4
Transformer rating	123.7	E3	58	E4	112.5	E3	60	E4
Customers' PVs	62.5	E3	32.5	E4	57.5	E3	40	E4
Roof-space PVs	14	E3	6.6	E4	12.8	E3	6.8	E4

(**c**)

Loading Level	100% of Transformer Rating				50% of Transformer Rating			
PV Connection	Balanced		Unbalanced		Balanced		Unbalanced	
HC References	HC	Limit	HC	Limit	HC	Limit	HC	Limit
Peak load	122	E3	40.7	E5, E4	221	E3	81.5	E5, E4
Energy consumption	98	E3	31.8	E5, E4	170.9	E3	64.8	E5, E4
Transformer rating	122	E3	40.7	E5, E4	110.5	E3	40.7	E5, E4
Customers' PVs	83	E3	26	E5	76	E3	26	E5
Roof-space PVs	9	E3	3	E5, E4	8.4	E3	3.1	E5, E4

The HC values w.r.t peak load and the energy consumption were greatly dependent on the peak load of the network, as depicted in Equations (2) and (3), respectively. Thus, scaling the peak load w.r.t transformer rating generally impacted the HC w.r.t these two references without significantly impacting the other reference definitions.

$$HC_{(\text{Peak Load})} = \frac{\text{Peak PV value}}{\textit{Peak Feeder Load}} \tag{2}$$

$$HC_{(\text{Energy Consumption})} = \frac{\text{Total Yearly PV Production}}{\textit{Annual Energy Consumption}} \tag{3}$$

The denominator in Equation (3) is calculated by aggregating the hourly loads of the network for one complete year and thus scaling the maximum load of the network against the transformer rating changes this value for two loading levels. The HC analysis of this work shows that HC of the same network calculated considering different references varied considerably with the HC w.r.t peak load, giving the highest numerical value among all the HC references.

HC of the balanced PV deployment case was higher than the unbalanced PV case for all the regions. In the case of a balanced PV connection, the HC was primarily limited by the voltage violations in the rural region, and the transformer overloading (E3) became the major limiting factor in the case of suburban and urban regions, as shown in Table 5. The unbalanced PV scenario presented negative sequence voltage unbalance (E4) as the main limiting factor for integrating a large amount of PV in the LV system for rural and suburban regions. However, the neutral wire ampacity (E5) remained the dominating limiting factor with the calculation of the HC of urban areas for an unbalanced PV deployment scenario. Moreover, the overvoltage limit did not prove to be as restricting in the case of unbalanced PV scenarios in rural regions where the most dominant limiting factor was observed as negative sequence voltage unbalance (E4). Additionally, transformer overloading (E3) appeared as the major limiting constraint for the PV penetration in the case of a balanced PV scenario for 100% and 50% loading levels in the urban region. Figure 4 shows the average HC at two loading levels by averaging all of the percentage estimates in Table 6. It shows that the HC at 50% loading level of the transformer rating outscored the HC value at 100% loading level of the transformer rating.

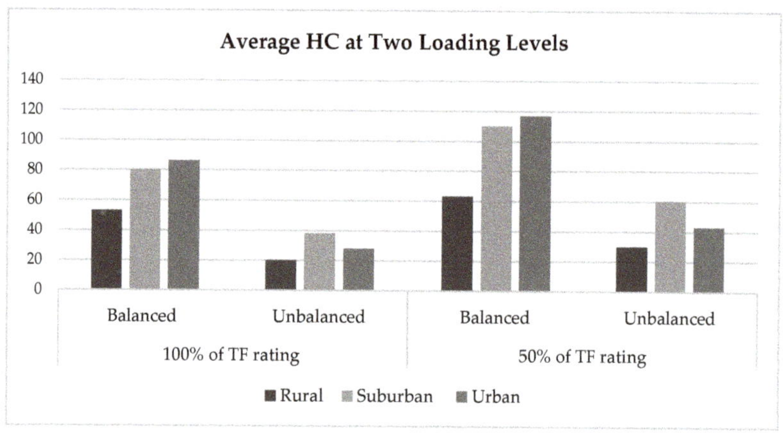

Figure 4. The average HC of three regions at different loading levels by averaging all percentage estimates of different reference definitions.

Overall, the voltage violation was observed as a key limiting factor for increased PV integration in the rural areas, and the transformer overloading noticeably limited the PV integration in the suburban and urban regions, as shown in Figure 5. The negative sequence voltage unbalance appeared as the limiting factor in all three regions, while the neutral wire ampacity limited HC of the urban region.

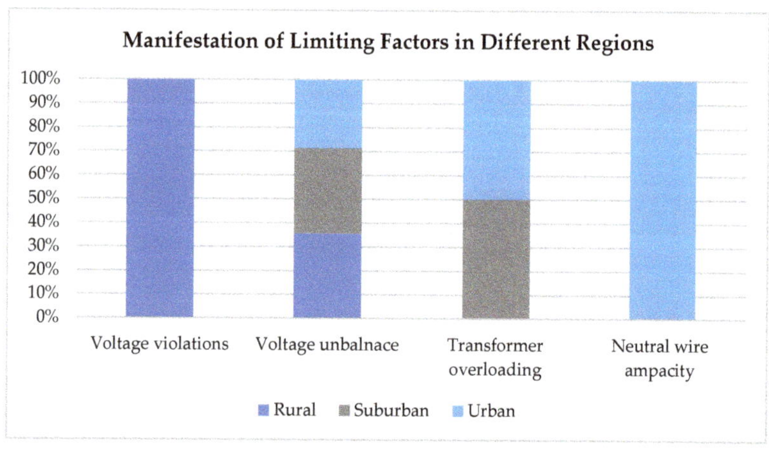

Figure 5. Quantitative analysis of HC limiting constraints for the investigated cases in Section 4.1.2.

Finally, the HC values of the networks were investigated at different load power factor values of 0.95, 0.8, and 0.7, with research results showing that the HC values remained almost consistent most of the time by changing the load power factor. There were slight fluctuations in the HC values that can be accredited to different load selection from the pool for each analysis. The results of this analysis are shown in Table 7 and also Figure 6.

Table 7. The HC variation based on changing load power factor for 100% and 50% loading levels for balanced PVs in rural, suburban, and urban regions.

Region	100% Loading Level			50% Loading Level		
Peak load	74	75.7	78	120	123.7	130
Energy	45	46.6	50	74	76.9	82
TF rating	74	75.7	78	60	61.8	65
Customers	62.5	62.5	62.5	50	50	50
Roofspace	10.58	10.8	11	8.6	8.8	9.3
Suburban	**P.F = 0.95**	**P.F = 0.8**	**P.F = 0.7**	**P.F = 0.95**	**P.F = 0.8**	**P.F = 0.7**
Peak load	123.7	122	120	225	224	224
Energy	80	73	74	140.5	145	144
TF rating	123.7	122	120	112.5	112	112
Customers	62.5	62.5	62.5	57.5	57.5	57.5
Roofspace	14	13.9	13.7	12.8	12.8	12.8
Urban	**P.F = 0.95**	**P.F = 0.8**	**P.F = 0.7**	**P.F = 0.95**	**P.F = 0.8**	**P.F = 0.7**
Peak load	122	122	119.5	221	221	221
Energy	98	97	94	170.9	178	163.8
TF rating	122	122	119.5	110.5	110	110.7
Customers	83	82	81	76	75.3	75.6
Roofspace	9	9.3	9	8.4	8.4	8.4

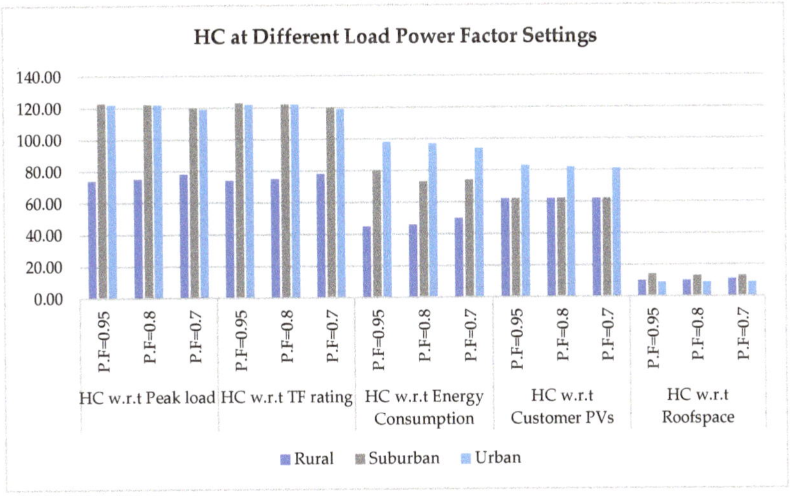

Figure 6. HC comparison for balanced PVs by taking maximum load as 100% of transformer rating for three load power factor settings of 0.95, 0.8, and 0.7 for the three regions.

5. Discussion

The comparison of HC w.r.t different references revealed the fact that HC is not a single value. Therefore, the HC must be carefully investigated in the context of the used references before integrating a large amount of PV generation in the network. The HC values varied considerably depending on used reference, technical limit values employed, share of single-phase PVs, the number of customers to be served, loading level, PV deployment, and location criteria. The main focus of this article was to compare the PV HC of the same networks by using different HC definitions as a reference. The test networks (rural, suburban, and urban) of this study were formulated in [2] by utilizing the real load consumption data of different customers characterized by their unique heating modes. A MC simulation approach is beneficial for simulating a high variance of networks as we used randomly sampled load profiles of each type of customer (differentiated by heating modes) for a total year. The proposed MC model enabled to use the input data of the test networks in the form of real DSO (distribution system operator) survey on measured load data and the average network size as formulated in [2].

Heating is the major part of the electricity consumption in cold climates. The investigated networks in this study were based on the real Finnish DSO surveys and real load consumption data of Finland. Therefore, this article was more focused on heating methods for defining the customer types as it took into account the conditions encountered in Finland: heating, generated networks, and PV arrays for Helsinki region. In heating-dominated load case, the loads are highest in winter time, whereas in summer, when the PV production is highest, the demand of residential customers is at its lowest. It has been established by the careful investigation of the same networks under the same loading conditions and same geographical location that the HC w.r.t different references produce quite disparate results. It was further observed that the HC value of the same PV generation amount w.r.t peak load showed the highest numerical value, and HC w.r.t roof space resulted in the lowest numerical value, thus maintaining it as the most conservative reference definition of HC in this analysis.

The peak load is the most widely used HC reference throughout the literature; however, the HC w.r.t peak load depends on the load sampling and scale of the peak load w.r.t transformer rating. Therefore, the frequent load variations of the network resulted in inconsistent HC values. The HC w.r.t transformer rating showed almost similar values for each load sampling and thus proved to be useful for the HC assessment, as already proved by the research results of [14]. Moreover, it was observed that scaling the maximum load w.r.t transformer rating mainly impacted the HC values w.r.t peak load and energy consumption, which involve the peak load directly in the HC calculation. A quantitative analysis of the HC limiting factors strengthened the fact that the voltage violations mainly limit the HC of rural networks, and transformer overloading restricts the HC of urban and suburban regions. This observation supports the use of transformer capacity as the main HC reference. If the HC is clearly lower than 100% of transformer rating, the HC limitations are likely to be due to voltage issues. On the other hand, the scaling w.r.t customer PVs or roofspace may reveal the unused PV generation potential among the local community.

The scope of this article was to compare the HC with respect to different reference definitions without utilizing any means to increase the HC of the networks. However, the potential of storage devices, reactive power control, voltage control, inverter oversizing, and dynamic loading of components should be addressed, taking into account different HC references.

Author Contributions: Conceptualization, S.F., V.P., and M.L.; methodology, S.F., V.P., and M.L.; software S.F. and V.P.; validation, S.F., V.P., and M.L.; formal analysis, S.F.; investigation, S.F. and V.P.; resources, M.L.; data curation, V.P.; writing—original draft preparation, S.F.; writing—review and editing, V.P. and M.L.; visualization, S.F.; supervision, M.L.; project administration, M.L.; funding acquisition, M.L. All authors have read and agreed to the published version of the manuscript.

Funding: This research was funded by BusinessFinland SolarX project.

Institutional Review Board Statement: No humans or animals were involved in this study.

Informed Consent Statement: No humans were involved in this study. The metered consumption data used for the study was anonymous and did not include any personal information.

Data Availability Statement: The data used for the study is confidential and was submitted by the DSOs for the research project in question only.

Conflicts of Interest: The authors declare no conflict of interest.

References

1. Fatima, S.; Püvi, V.; Lehtonen, M. Review on the PV Hosting Capacity in Distribution Networks. *Energies* **2020**, *13*, 4756. [CrossRef]
2. Arshad, A.; Lindner, M.; Lehtonen, M. An Analysis of Photo-Voltaic Hosting Capacity in Finnish Low Voltage Distribution Networks. *Energies* **2017**, *10*, 1702. [CrossRef]
3. Torquato, R.; Salles, D.; Pereira, C.O.; Meira, P.C.M.; Freitas, W. A Comprehensive Assessment of PV Hosting Capacity on Low-Voltage Distribution Systems. *IEEE Trans. Power Deliv.* **2018**, *33*, 1002–1012. [CrossRef]
4. Weisshaupt, M.J.; Schlatter, B.; Korba, P.; Kaffe, E.; Kienzle, F. Evaluation of Measures to Operate Urban Low Voltage Grids Considering Future PV Expansion. *IFAC-PapersOnLine* **2016**, *49*, 336–341. [CrossRef]
5. Heinrich, C.; Fortenbacher, P.; Fuchs, A.; Andersson, G. PV-integration strategies for low voltage networks. In Proceedings of the 2016 IEEE International Energy Conference (ENERGYCON), Leuven, Belgium, 4–8 April 2016; Volume 2. [CrossRef]
6. Procopiou, A.T.; Ochoa, L.F. Voltage Control in PV-Rich LV Networks without Remote Monitoring. *IEEE Trans. Power Syst.* **2017**, *32*, 1224–1236. [CrossRef]
7. Bollen, M.H.J.; Rönnberg, S.K. Hosting Capacity of the Power Grid for Renewable Electricity Production and New Large Consumption Equipment. *Energies* **2017**, *10*, 1325. [CrossRef]
8. Hong, T.; Lee, M.; Koo, C.; Kim, J.; Jeong, K. Estimation of the Available Rooftop Area for Installing the Rooftop Solar Photovoltaic (PV) System by Analyzing the Building Shadow Using Hillshade Analysis. *Energy Procedia* **2016**, *88*, 408–413. [CrossRef]
9. Sorsanen, J. Assessment of Realistic Solar Electricity Production Potential of Grid-Connected Photovoltaic Systems in Tapiola District; Evaluation of Realistic Photovoltaic Production Potential for Grid-Connected Photovoltaic Systems in the Tapiola Area. Master´s Thesis, Aalto University School of Electrical Engineering, Aalto University, Espoo, Finland, 2015.
10. Rahman, F.F.S.; Adi, K.W.; Sarjiya; Putranto, L.M. Study on Photovoltaic Hosting in Yogyakarta Electric Distribution Network. In Proceedings of the 2018 5th International Conference on Information Technology, Computer, and Electrical Engineering (ICITACEE), Semarang, Indonesia, 27–28 September 2018; pp. 240–244. [CrossRef]
11. Alam, M.J.E.; Muttaqi, K.M.; Sutanto, D. Distributed energy storage for mitigation of voltage-rise impact caused by rooftop solar PV. In Proceedings of the 2012 IEEE Power and Energy Society General Meeting, San Diego, CA, USA, 22–26 July 2012; pp. 1–8.
12. Ekstrom, J.; Koivisto, M.; Millar, R.J.; Mellin, I.; Lehtonen, M. A statistical approach for hourly photovoltaic power generation modeling with generation locations without measured data. *Sol. Energy* **2016**, *132*, 173–187. [CrossRef]
13. Hoke, A.; Butler, R.; Hambrick, J.; Kroposki, B. Steady-State Analysis of Maximum Photovoltaic Penetration Levels on Typical Distribution Feeders. *IEEE Trans. Sustain. Energy* **2013**, *4*, 350–357. [CrossRef]
14. Arshad, A.; Püvi, V.; Lehtonen, M. Monte Carlo-Based Comprehensive Assessment of PV Hosting Capacity and Energy Storage Impact in Realistic Finnish Low-Voltage Networks. *Energies* **2018**, *11*, 1467. [CrossRef]
15. Atmaja, W.Y.; Sarjiya; Putranto, L.M.; Santoso, S. Rooftop Photovoltaic Hosting Capacity Assessment: A Case Study of Rural Distribution Grids in Yogyakarta, Indonesia. In Proceedings of the 2019 International Conference on Electrical Engineering and Informatics (ICEEI), Bandung, Indonesia, 9–10 July 2019; pp. 448–453. [CrossRef]
16. Vimpari, J.; Junnila, S. Estimating the diffusion of rooftop PVs: A real estate economics perspective. *Energy* **2019**, *172*, 1087–1097. [CrossRef]

MDPI

St. Alban-Anlage 66

4052 Basel

Switzerland

Tel. +41 61 683 77 34

Fax +41 61 302 89 18

www.mdpi.com

Clean Technologies Editorial Office

E-mail: cleantechnol@mdpi.com

www.mdpi.com/journal/cleantechnol